A Digital Liberia

A Digital Liberia

How Electrons, Information, and Market Forces Will Determine Liberia's Future

Darren Wilkins

iUniverse, Inc.
New York Bloomington

A Digital Liberia
How Electrons, Information, and Market Forces Will Determine Liberia's Future

Copyright © 2010 by Darren Wilkins

Edited by: Miyesha Cheeks and Jennifer Wilkins

All rights reserved. No part of this book may be used or reproduced by any means, graphic, electronic, or mechanical, including photocopying, recording, taping or by any information storage retrieval system without the written permission of the publisher except in the case of brief quotations embodied in critical articles and reviews.

The views expressed in this work are solely those of the author and do not necessarily reflect the views of the publisher, and the publisher hereby disclaims any responsibility for them.

iUniverse books may be ordered through booksellers or by contacting:

iUniverse
1663 Liberty Drive
Bloomington, IN 47403
www.iuniverse.com
1-800-Authors (1-800-288-4677)

Because of the dynamic nature of the Internet, any Web addresses or links contained in this book may have changed since publication and may no longer be valid.

ISBN: 9781450258722 (sc)
ISBN: 9781450258777 (dj)
ISBN: 9781450258760 (ebk)

Printed in the United States of America

iUniverse rev. date: 10/25/2010

Dedication

To the great people of Liberia, who have long been on the other side of the digital divide

In memory of my father, Jerry I. Wilkins Sr., who didn't live to see me accomplish all that he had asked me to accomplish

To my mother, Vor-younoh E. Wilkins, who never sat in a classroom but was always there to help me do my homework

To Geraldine Daisy Wilkins, my sister, who paid for my first formal computer education

To Miss Miyesha Cheeks, for standing with me and being there when I most needed her

To the old and new generation of the Wilkins family and all the other families in Liberia and other developing countries whose future will be impacted by technology

Epigraph

Our vision is for Liberia to become a globally competitive knowledge and information society where lasting improvement in social, economic, and cultural development is achieved through effective use of ICT.

–*President Ellen Johnson Sirleaf*

CONTENTS

FOREWORD

I am delighted at the invitation to write a foreword to this book, not because I think it is the *final* word on information and communication technologies (ICTs) in contemporary Liberia, but because I think it sits as one of the *first* words, and certainly the first book-length treatment, of what will hopefully become a growing and expanding literature. In this valuable contribution, Darren Wilkins describes how computers and communications can assist Liberia in building its education, government, business, security, health and agricultural capabilities.

But what really are the prospects for ICTs to serve as a critical tool in Liberia's development? On the one hand, the promise appears significant: The communications sector is vibrant with phone subscriptions enjoying one of the world's leading rates of compound annual growth according to the International Telecommunications Union (2010); and usage costs are reportedly the lowest in West Africa (Southwood 2007). But on the other hand there are challenges: ICTs have yet to realize their full promise in remaking business, education and government as Wilkins makes abundantly clear in this text; indeed most schools languish without any computers, most government offices are yet to be robustly connected, at times leadership has faltered, and only a tiny proportion of Liberians have ever been on the Internet.

Indeed, if Liberia is to harness the power of ICTs in her national development she will need to employ an all-hands strategy, bringing together the energy and know-how of a broad set of stakeholders. This includes Liberians who have remained in-country throughout the war years, more recent returnees, Liberians in the Diaspora, regional and international experts, and the full range of development partners. In today's globalized ICT sector nobody can go it alone, and this is all the more true for conflict-affected countries like Liberia.

This is one reason that Wilkins' contributions, even beyond this book, demand our attentions. He represents an engaged and committed actor from the Diaspora. Here he describes on-going work at the BW Harris High School; every year he travels to Liberia and volunteers his time and money at his alma mater. The fruit of this labor is a school now equipped with a modern computer lab. And Wilkins, I am told, is one of the very few Liberians in the Diaspora to have provided expert comments to the Ministry of Post and Telecommunications on the draft ICT national policy documents.

If Wilkins represents an engaged Liberian in the Diaspora then I, most immodestly, count myself as an expert from the outside. My general fascination with and focus on ICTs echoes Wilkins' early experiences; I too came to computers as just a young (and admittedly nerdy) boy. But my engagement with Liberia, as an outsider, is more recent and necessarily different from Wilkins'. I've been working as an external partner researching the role of computers and communications in international development (ICT4D) for more than a decade now. This includes numerous engagements across Sub-Saharan Africa including most significantly Nigeria, Rwanda, Ethiopia, Kenya and Ghana. Perhaps I have some reputation across parts of Africa for work in this space and so it happened that in 2006 the George Soros Foundation Network approached me, asking if I would do a simple ICT sector assessment of Liberia. The result (Best, *et al.* 2007) forms the first broad analysis of post-conflict Liberia's state of ICT readiness. But what I imagined would be just one simple study was instead the start down a slippery slope as I increasingly found myself collaborating on more and more work in Liberia. Now going on five years after first stepping into Liberia, I continue with some engagements there, look forward to even more on-going interactions, and view with extraordinary pleasure the steady progress made across many areas of development in Liberia.

It is with this set of experiences that I return to a reflection upon this book. Through more than 250 pages we are given plenty to think about – much of it I agree with, some of it I might dispute, all of it

I find of interest. But for me, just *what* is said in this book is often secondary to *why* it is being said. In his Preface, Wilkins answers for us this question as to why: "What follows is a modest attempt to encourage Liberians to… build what they want to be and not what they used to be.… This is a part of my contribution to Liberia's future." Indeed. In this extended family of people who care deeply for Liberia, a family I count myself a member of, I would hope that we all can make a similar scale of contribution.

Dr. Michael L. Best
Assistant Professor
Sam Nunn School of International Affairs and the School of Interactive Computing
Georgia Institute of Technology

Preface

For someone with extravagant taste, it would seem logical that I would be most interested in making money from this book; but that is not the case. After years of academic and independent research, I thought that by combining my published writings I would give others the chance to engage in the debate of how technology will impact Liberia's future. As a member of Generation X, I often read and reflect on the history of Liberia and the things that our forefathers did or did not do. Then I compare my generation and the changes we have experienced to that of my forefathers and the changes that they experienced, and I appreciate that both are now considered things of the past. That realization, coupled with Liberia's history, has given me a vision that many of my contemporaries sometimes argue is a mere dream. But many of those who had argued against the changes that I foresaw were silenced when those changes were implemented to the benefit of the Liberian people. In fact, the advent and implementation of the ATM machine in Liberia is a quintessential example of what I just referred to. I remember vividly that when I first mentioned that ATM machines will be used in Liberia, a friend of mine was quick to point out that it would be impossible to have ATM machines in Liberia considering the infrastructure, security, and culture of the country. Today, ATM machines can be found in several locations in Liberia!

The information age is upon us, and the changes that it brings are irreversible. If Liberians choose to ignore these changes, she will do so at her own expense. But, knowing Liberians, adapting to the technological age will not be a problem. All that is needed is awareness, will, and good leadership. Therefore, I am writing this book, not only to spark a debate about the ensuing changes in Liberia or create awareness, but as my contribution to the process leading to a digital Liberia. It is my hope that everyone, including the most recalcitrant pessimist, will use it as a guide as Liberia gravitates toward the information age.

Throughout this book, I refer to events that I believe can and will happen as we move ahead, and I make reference to the past and the present to relate to the future. Some of the things that I predict now may perhaps seem impractical and obscure but will eventually occur; and with ubiquity. And when they do, if I am lucky, I may just be recognized for being farsighted. Conversely, other ideas that might not materialize may be used by pessimists and cynics to prove that I was simply living in a dream world. But isn't that always the case? Nobody fathomed that a person in Zwedru, Liberia, would use a cellular phone to call Minneapolis, Minnesota. Nobody expected to type a letter and make changes without using correction fluid. But it is happening! Based on these events, I grew up with great optimism and farsightedness. And as I see it, Liberia has a brighter future than what many of its people may envision. Liberians have to be optimistic, explore the unfamiliar without fear, and be persistent. Liberia will only get better and be prosperous.

I must make another confession. All through this book, I inject my personal life experiences to illustrate my points. However, this book is not intended to be my autobiography; it is about Liberia's future. I hope that the book is perceived and analyzed as a guide toward a debate that will lead to a digital Liberia.

This book is not only intended for those in the fields of computing and information and communications technology, but rather for every stakeholder in Liberia. I fervently hope that everyone joins in this debate to garner understanding and emerge with ideas that can lead to new innovations. Liberians, like all other citizens of the world, have immense aptitude. But unlike the rest of the world, Liberians' aptitude has been suppressed by slavery, dictatorship, starvation, poverty, war, corruption, and above all, pessimism.

Finally, I was tempted to write a more technical book or a book on a certain technology or technologies. But I decided that that could be done some time in the future; hence, this book. I knew that had

I written a book strictly about computers or computing, the debate would have been limited and geared toward a particular portion of the population. And that is not my intention. Although there are some parts of the book where it seemed inevitable to include information and communications technology (ICT) lingo or jargons, I tried my best to be as nontechnical as I could be.

What follows is a modest attempt to encourage Liberians to explore what is to be anticipated and give them the basis on which to build what they want to be and not what they used to be. It is also an effort to encourage Liberians to dare to hope and to achieve those things that their forefathers never thought would be possible. This is a part of my contribution to Liberia's future.

ACKNOWLEDGMENTS

It's hard to acknowledge everyone you have been helped by, especially when there were so many people involved. But I will take a stab, and I hope that those whose names I inadvertently omit will forgive me for doing so. I can assure them that I am most grateful for all the help they've given over the years.

I would like to acknowledge my family: Mom, Geraldine, Jennifer, Archibald, Maribel, Abigail, Glen, Virginia, Fenell, Archelle, Darrean, Ashley, Dominique, Paco, Murphy, Lolita, Raheem, Little Geraldine, Xavier, Micah, Jareem, Bobby , Jerry, Larry, Marvelyn (Teta), Pamela, Kevin, Diamond, Rosaline, Stanley-Godspower, Chuke, Kester, Jademul, Mac, Teta, Zayiea, James, Albert, et al.

Thanks to Mrs. Beti Wilkins, for the initial support and for previously enduring my countless excuses for not completing domestic tasks, as well as my constant preoccupations and frequent absences.

A big thank you to the following individuals for inspiring and motivating me to write this book: Herman Williams Getahn Ward, James Buchanon, Macsu Hill, Debbie Bovee, Dr. Rowena Kelley, and the entire class of "88" of B. W. Harris High School.

Special thanks Mrs. Vivian M. Smith and Mrs. Puchy Franco, to whom I shall always be indebted and who proved to me that sometimes the person you least expect to be a good friend can actually be a best friend.

Thanks to Mr. Handel Diggs, Jimmy Neal, Anthony Deline, and the entire B. W. Harris Alumni Association and the B. W. Harris

School in Liberia, for the cooperation and the opportunity given me to implement technology at the school. The implementation of technology at the school was an experiment that demonstrated that the potential for Liberia to change into a digital Liberia does indeed exist.

Thanks to the management and staff of *Liberia Daily* for giving me the opportunity to bring ICT awareness to the Liberian society through weekly articles. A big thank you to Mr. Kenneth Best, Miss Boto Best, Mr. Bai Best, and Mr. Alaska Johnson for closely working with me. Alaska, I still owe you the $40.00 that you spent on the souvenir paper you purchased for me in December 2008.

I cannot complete this acknowledgement section if I do not express thanks to two people who have a special place in my heart. They are my sister Jennifer Wilkins and Miss Miyesha Cheeks. These two ladies spent countless hours reading and editing the draft of this book. A very special thanks to Miyesha for staying up all night and enduring my boring conversations over the phone, e-mails, and Skype.

Finally, I would like to thank all of those whose name I did not mention but contributed toward the success of this book. Our next task is to work toward the building of a digital Liberia.

INTRODUCTION

When technology began to infiltrate and demonstrably impact Liberian culture, I had already immigrated to the United States toward an industry with great potential to effect change in Liberia for generations to come. What I recall about Liberia before I left was that there was limited technology available and there were definitely no discussions about it and the Internet—at least to me and those whom I knew. The closest I came to glimpsing technologies that seemed to indicate what was to come was when compatriots who lived or visited developed countries returned with gadgets in hand or stories to tell.

My journey into the world of technology began when I was an eleven-year-old student in the fifth grade at B. W. Harris School nestled at the ascent of Snapper Hill on Broad Street in Monrovia. I, along with my best friends Victor Fares, Robert Silla, and Getahn Ward, had an insatiable appetite for computer games—Pac-Man, to be specific. Each one of us had our own portable Pac-Man device and spent countless hours playing and mastering the game. Later that year, an entertainment center known as Monte Carlo Amusement Center opened and housed larger video arcade games that required coins to play. My first experience with arcade games at Monte Carlo sparked my imagination and thoughts of building my own computer games … of course, that was a dream then, and I was only eleven!

Seven years later, during the second semester of my freshman year at the University of Liberia, I dropped two of my three classes to pursue a different path. At that time, I was majoring in civil engineering, but

I believed that my passion lay elsewhere—something in computers and not in economics, as my father had hoped. Hence, without any money, I sought to get answers regarding enrollment at a computer school I had learned of while riding in a taxi cab.

When I arrived at the school, named Laubtubdina Computer Inc. (LTD), I learned that the cost of the program was $1,500 Liberian dollars, which I did not have. However, I learned that the school was owned by Mr. Lawrence Dwamina (deceased), an alumnus of my alma mater, B. W. Harris School. I hoped that connection would earn me some sort of relief in terms of making tuition payments. But that was not the case!

There at LTD, my interest in computers was unique. While others sought to garner a certificate for employment purposes, I was in the program primarily because I wanted to learn the "magic" that made computers work. Unfortunately, the courses that were taught there were not challenging enough to satisfy my curiosity. The courses included MS-DOS, WordPerfect, Lotus 123, and dBase. As an indication of the little challenge they posed, I consistently scored above the targeted point on every test—130 percent or 115 percent, when the target score was 100 percent. The extra points were available to help students who struggled on certain parts of the test.

My quest to acquire in-depth knowledge of computers led me to join in on conversations had by the former owner of the school. Through him, I learned of BASIC and C programming. Equipped with this knowledge and knowing that it would take an act of God to have LTD teach me programming, I sought other avenues that could provide me with information on computer programming. Disappointingly, my quest led me to the University of Liberia's library; at that time it did not have a single book on computer programming.

The most defining moment of my journey into the world of computing was the day I purchased a book that showed a photo of a nondescript guy named William Gates and his company called Microsoft. The book discussed programming languages such as FORTRAN, COBOL, BASIC, etc. My attention was arrested by the subheading that read *Programming in BASIC*. As I flipped through the pages, I soon realized that it was an introduction to BASIC, which was good enough for me at that time. As I read through the book, I went back to LTD—even though I had completed my course before schedule I was still allowed in because classes were still in session. Clandestinely, I used that time to figure out the computers while referring to the book that I had bought … and no one knew!

My experience with my "book" and LTD did not last too long, because the session had ended and I did not have enough money to enroll in a follow-up course; that seemed to be the end of my computer studies. But I kept my book and referred to it occasionally in hopes that I would someday have another shot at a computer, at which time I would harness my skills for future use. But what followed after that little episode was a bloody civil war that brought more destruction to the country than its citizens had ever imagined. This was in the latter part of 1989. I do not believe that I will ever forget that time!!

In 1991, I was hired by Petrol-1 Inc., initially as the assistant technical manager. My managing director, Mr. Elias Karaan, encouraged me to learn everything about the company, including the business model. My job was to work with my immediate manager, Mr. Jorneh Tarpeh, to perform petroleum inspection, control, and distribution. I was able to quickly grasp new knowledge, expand on it, and adopt a paradigm of finding new, innovative, and exciting ways to do old things; thus my successful tenure with the company.

In 1997, when I arrived in Lehigh Acres, Florida, I had the opportunity to gain full access to computers, as my brother-in-law, who had attended a vocational school studying computer maintenance and hardware, had left a stack sitting at the house. It appeared that Winston (White Jr.), my brother-in-law, was more of a "hardware guy," because there were computer parts strewn about the house. But thanks to him, I learned all the parts of a computer before I entered a classroom. At the time, the operating system in use was Windows 3.1, and it was my first time experiencing a computer without the DOS black and white or black and gold interface. This new interface was very attractive and pulled me away from many domestic duties for several days.

As the days went by, I managed, out of curiosity, to open one of the computers; Winston had told me that the only way I'd learn about computers was to delve in completely and boldly. I managed to disassemble and reassemble the computer with ease; I was very good at doing things like that. When it was time to load the operating system, the computer would not turn on. It seemed that I had succeeded in reassembling the computer but had assembled something incorrectly. So I opened it again and realized that I had forgotten to insert the memory chip.

In the fall of 1998, I decided to enter Edison State College (previously Edison Community College) in pursuit of higher education. This marked my first experience at an institution of higher learning in America. My first semester was a challenge for me as I had just arrived in the country and was being charged an exorbitant amount as an out-of-state student. Not only did I have to pay tuition that I could not afford, but I was required to have my transcript from the University of Liberia evaluated by an educational evaluator in Miami, Florida. What was more frustrating was that after my transcript had been evaluated, I had to repeat several classes that I had taken at the University of Liberia. This was a pivotal moment in all of my academic life, as it influenced the decisions that led to the journey on which I have embarked.

Following the receipt of my evaluation and after meeting with an advisor at Edison State College, I was given a course catalog and advised to take certain courses that would give me "a feel" for the American higher-educational system. The first three courses were statistics, student learning skills, and business communications. I had registered to pursue an associate's degree in computer science. But being charged out-of-state tuition and with little income to pay for my classes, I could not take all three of the classes required that semester. I also could not get financial aid because I had just come to the country. Also, the little that I earned working a full-time job had to be saved to bring my parents to the United States, because my father desperately needed medical attention. This would happen in 1999.

Hence I was constrained to take only one class because I was paying the tuition out of pocket. I ended up taking business communications because I felt that I needed to be able to communicate well with the people whom I would have to interact with.

After registering for the class and being so excited about the thought of attending school, I proceeded to the library to purchase my textbooks, other school supplies, and a sweater that read "Edison Community College, Visit our Web site at: www.Edison.edu." The fact that the sweater had "Edison Community College" written on it was all

I cared about. At that time, I was oblivious to what a Web site was or what www.Edison.edu meant because I had never seen or heard of the Internet. I had planned to wear the sweater and take a photo of me in it to send to friends in Liberia. I never did take that photo. I wore the sweater only once that fall, on a Tuesday, which happened to be the first day I attended my business communications class and also marked the beginning of my academic career in the United States.

The first day I entered Edison College, I was very excited; even more than I had anticipated. It was a new experience with new people in a new environment. I took a seat in the middle of the class adjacent to another student whom I later came to know as Christopher P. That day started a friendship that would last for many years; even as we became professionals. Chris and I would subsequently spend much of our time in the college's computer lab. By then, I was already familiar with computers, although not as familiar as my classmates were.

After a series of lectures and exploring Microsoft PowerPoint, our instructor stated that it was time to move on to the "information superhighway," a term I had absolutely no clue about and one that I later came to understand meant the Internet.

As we began our exercises, I received instructions from my instructor to click on the icon that looked like an "N," which I did very eagerly. A page came up that said "Welcome to Edison College." On it was a textbox labeled "Search," which I hastily clicked as a result of my eagerness. The next instruction from my instructor was a request to type in a keyword in the search textbox and hit the ENTER key. She wanted us to research a topic and then create a presentation using MS PowerPoint 95. To me that was a "call to arms," because the first keyword word that I typed was "Liberia" and when I did, several links populated, some of which had photos of the Liberian flag and the then Liberian president (Charles Taylor). That day marked my very first experience with the Internet.

What followed was a profound addiction to the Internet. It captivated my attention, and consequently I missed several lectures and shirked many of my other responsibilities. I realized that my newfound love for the Internet was the genesis of a future initiative; I just did not know what it would be. I did, however, know that whatever that initiative would be, it would heavily gravitate toward Liberia.

The next few years would be successive quests for knowledge in ICT at Hodges University (formerly International College) and various media, including professional development initiatives at my places of work. Those quests led to a number of college diplomas, ICT certifications, ICT conferences, trainings, and alignment with several ICT-related organizations, including: International Standards for Technology in Education (ISTE), *Association for Computing Machinery (ACM)*, Computer Science Teachers Association (CTSA), Southwest Florida Linux Users(SWFLUG), Atlanta Linux Enthusiasts (ALE), AWESome Atlanta (Cloud Computing User's Group), UN Online Volunteers (UNOV), to name a few. Hodges University Graduate School of Technology was very interesting and exciting. I honestly believed that it was there that I developed a broader perspective of ICT, especially as it related to underdeveloped countries. It was also during my graduate school years that I accomplished the first published academic work; project WINWILE.

Project WINWILE, or Windows interoperability with Linux in the enterprise, was my graduate project, which involved interoperability of disparate systems. Three operating systems were used in that project: Windows XP, Linux (Red Hat), and Macintosh OS 9. At the time (2004) there were limited capabilities in the area of interoperability between disparate systems. XML and other languages and protocols were still not popular. With Samba protocol and WINE (www.winehq.com), I managed to create an environment that allowed all three of these systems to work. Upon completion of the project I wrote a paper that was presented at a computing conference in South Carolina and subsequently published in the *Association for Computing Machinery(ACM)*. The paper lists the names of my advisors and professors as co-writers.

Graduate school at Hodges University involved many academic projects that could have been turned into solutions for the business environment. But while I had the will and ideas to invent new solutions for economic development, initial funding and resources were not available to pursue such endeavors. For this reason, those projects were neglected in favor of work and other pursuits. I have always felt that if I had the initial capital to further proceed with one of my ideas, I could have achieved some level of success. And even if that success did not measure up to the likes of Bill Gates or Michael Dell, I would have accomplished something that would have given me some kind of national recognition for bringing change in the ICT world.

There are two major projects that would have projected me into the spectrum of ICT innovators during the early and mid-2000s. The first attempt was the creation of my own computers to be sold to the market. At the time, my target audience was underprivileged families, immigrants, and all those who could not afford to buy new and expensive computers. I began refurbishing computers that I purchased from thrift stores and began reselling them for little money. Initially, this endeavor was quite successful to the extent that I decided to design my own form factor or computer case with my name and logo. I also offered free Internet access provided by NetZero, which distributed its software free of charge as a marketing promotion. However, none of this materialized as I found myself struggling with business decisions. It is one thing to be a visionary computer guy; it's another thing to be a shrewd businessman. I was the former but was far from the latter. And so I decided to continue my regular job and head back to school.

The other project that could have gained some traction had I not detoured was the creation of a laptop with a built-in projector. I noticed that while traveling, sales representatives always carried their own projectors and laptops in the event that their final destination was not equipped with this technology. Hence, building a laptop with a built-in projector would have reduced the load sales representatives carried. While I have yet to see this kind of device on the market, I have no doubt that it will soon find its way to stores. And since I did not listen to friends who had advised me to have the idea patented, I have no claim over such an idea other than the paper I wrote while in college and the mere mention of it in this book.

My professional life has also had an impact on this journey. Having worked in both corporate America and for government, I do not notice much difference between the two. Most of my professional experiences were garnered from my tenure with the Lee County School District, where I served as a technology specialist. In the business sector, I worked with the American Power Conversion Company, Sony Electronics, Pall AeroPower Corporation, and as a consultant with General Electric. Each of these entities provided me with unique experiences that will take me more than a book to discuss. But my time spent at those companies and the Lee County School District allowed me to experience every aspect of information technology, including databases, programming, Web development, networking, Internetworking, computer telephony, electronics, call centers, etc. I owe a debt of gratitude to my former employers for exposing me to a plethora of ICT skills and knowledge.

In my personal and non-work-related endeavors, I wandered into the area of open source software, e-commerce, submarine fiber-optic cables in Africa, and cloud computing. You see, I have always believed that knowledge should be shared; that as long as we share knowledge, the world will be better place. I found sanctuary in the open source community because it allowed me to harness my intellectual capabilities for the betterment of society. In the open source community, collaboration is mundane, and because I favor collective problem solving, this seemed the ideal way to accomplish some of my ambitious goals. During this time, I managed to engage in a lot of activities in several communities, including the Ubuntu Linux community. I also engaged in some international initiatives that led me to Liberia, where I conceived, planned, designed, and implemented the Smart Technology Project that I discuss in a later chapter in this book.

Apart from what I have mentioned, there are other activities and initiatives that I have purposely chosen not to mention in this book, since it's my first one. But I hope to do so at some time in the future. For now, my goal of writing this book is to start a debate on ICT in Liberia. It is also my hope that other African and developing countries will use this book to start a similar debate in their countries. After all, there is one thing we all know: ICT can bring economic development and bridge the proverbial digital divide.

Part I

Origin

CHAPTER 1

A BRIEF HISTORY OF LIBERIA'S SITUATION

Liberia is a small country situated on the west coast of Africa. It measures forty-three thousand square miles, about the size of Tennessee, and shares borders with Cote d' Ivoire, Guinea, Sierra Leone, and the Atlantic Ocean. The Republic of Liberia was colonized by the American Colonization Society in response to America's issues with slavery and racial incompatibility. In July of 1847, Liberia won its independence and became the first free African republic. Its government was designed to resemble that of the United States. Joseph Jenkins Roberts, who was born in Norfolk, Virginia, served as Liberia's first president. Roberts had immigrated to Liberia in 1829, a few years after the freed slaves had settled in Liberia, and became president when the country gained its independence. Roberts served two terms as president of Liberia. The first term spanned 1847–1856, and as Liberia's seventh president his second term spanned 1872–1876. Most, if not all of the presidents that succeeded President Roberts shared one thing in common: they were all Americo-Liberians, descendants of freed slaves of the United States of America. This lasted until the fateful morning of April 12, 1980.

The morning of April 12, 1980, marked the end of Americo-Liberian True Whig Party totalitarian rule, which over time created a palpable division, evident by socioeconomic and ethnic disparities, between the indigenous and the Americo-Liberians. That morning Liberia experienced its first military coup d'etat, which brought the country's indigenous sons to the highest office of the land. My memory of that morning is still very vivid. Although a child, I can recall watching an intimidating group of soldiers protecting a young master sergeant who frenetically read a speech justifying the coup. The leader of the coup, Master Sergeant Samuel K. Doe, a twenty-eight-year-old soldier who, it was reported, had violently overthrown President William Tolbert, had the complete attention of a nation paralyzed by shock and fear. This assault was orchestrated with neither a comprehensive plan for Liberia then nor an understanding of the implications this coup would have on Liberia's future.

Ironically, Liberia experienced a modest level of economic development during the Doe regime. The entities that represented what is currently the telecommunications and ICT sector, the Liberia Telecommunications Corporation, Liberia Broadcasting System, and the Ministry of Postal Affairs—were all viable entities during Doe's regime. Analog telephone systems, electricity, and running water were available in certain areas. And the country's infrastructure was relatively intact at that time because President Doe had continued the infrastructural development initiatives started by President William R. Tolbert as well as those his government embarked upon. There was no Internet nor large data centers at the time, although banks and other institutions did have their own data-processing centers, obviously not like the ones we have today. There were no pagers, cellular phones, or smart phones, either. What Liberia did have in the form of technology were typewriters, the fax, telegraph machines, analog phones (land lines), and of course the regular postal system, which everyone relied upon. During those years, the government did not concern itself much with computer information technology, because computers were still in their infancy. While there were few Liberians who showed interest in and embarked upon computing careers, there were few if any opportunities or support for this field. In fact, universities in Liberia offered only introductory courses in computer science. Even today, Liberian universities have yet to introduce a robust computer science or information and communications technology program.

Despite all of the progress that Liberia had made in terms of development, it seemed on the verge of collapse due in part to the planned but forceful return of some Liberians who had defected to other countries after the coup. In December of 1989, it came as no surprise when Liberia was invaded by rebel forces of the National Patriotic Front of Liberia (NPFL), by way of the Côte d' Ivoire border. The NPFL was led by Charles Taylor, who subsequently proclaimed himself president. Taylor would be elected president in 1997 by the people of Liberia.

In 1990, during the incursion, Doe was assassinated by another group of rebels led by a self-proclaimed general known as Prince Y. Johnson. By this time, Liberia lay in ruins, the entire infrastructure destroyed, and although certain areas still had telephone service and the rebels began restoring services in their captured areas, Monrovia (the capital of Liberia) continued to crumble. The fighting lasted a few more months after the assassination of President Samuel K. Doe despite the rebels' claim that Doe's demise would bring an end to the war. By this time the world had seen enough and something needed to be done to preserve what was left of the country.

Sometime in 1990, the Economic Community of West African States (ECOWAS) intervened and helped negotiate a peace agreement among warring factions and what was left of the Doe government. As part of the intervention, ECOWAS also sent a Nigerian-led West African peacekeeping force, ECOMOG, to Monrovia, Liberia, which restored peace and calm and subsequently installed an interim government led by Dr. Amos Sawyer.

During Sawyer's interim administration the country experienced little in the way development, and unfortunately it was during this time that technology was poised to revolutionize the world. The personal computer was now being used as a business tool, the Internet had gone commercial, there was a change from Gopher to the World Wide Web, e-mail began to gain popularity and momentum, and there was Vice President Al Gore's "Gore Bill," or the High- Performance Computing and Communication Act of 1991 (California4gore n.d.). All of these developments occurred while Liberians were experiencing the wrath of the war. I often wonder what would have happened in Liberia had there been no civil war. Would Liberia have gotten on the information superhighway earlier? I would like to think yes!

In 1992, with military aid from Libya and Burkina Faso, Taylor's forces laid siege to Monrovia and engaged in fighting with ECOWAS forces, a fight that broke the relative peace that had been achieved and the new hopes of Liberians who had been totally devastated. Between 1993 and 1994 a number of cease-fires were established, but the warring factions, especially the National Patriotic Front of Liberia, refused to agree to a peaceful and amicable solution. After a long battle with the ECOWAS's monitoring group, ECOMOG, Mr. Taylor and his forces agreed to a cease-fire and the installation of an interim government inclusive of representatives from various warring factions. Surprisingly, what followed was a general election that ushered Mr. Charles Taylor in as the elected president of postwar Liberia.

On August 1, 1997, Charles Taylor became president of Liberia after the end of the "first" Liberian civil war. No sooner had Taylor taken office than he began to lose the favor and confidence of the masses, just like his predecessor, Samuel K. Doe. His government reportedly prolonged the suffering of the Liberian people as his followers continued to torment the very people he and his office had sworn to protect. Yet, despite Taylor's shortcomings, he did bring some degree of change to Liberia. It was during his administration that cellular phones were introduced to Liberia, bringing a major paradigm shift in the way Liberians communicate. The introduction of cell phones to Liberia's market was a great milestone. It has forever changed the way Liberian's live and bridged a gap that had existed for years between Liberia and the Western world.

In 2003, another disruption in the peace process led to yet another battle between newly formed warring factions and the Taylor-led government. President Taylor was forced to relinquish power to his vice president as a condition for peace. He was later made to leave Liberia for Nigeria, where he lived prior to his arrest and subsequent prosecution at The Hague. This was made possible through the intervention of the United Nations (UN) and the Organization of African Unity (OAU). After his departure a new day dawned, leading to the holding of free and fair elections. The presidential elections ushered in the first female president in African history, Mrs. Ellen Johnson Sirleaf, a Harvard-trained public administrator. Since then, the country has enjoyed relative peace and stability and has been making efforts to rebuild and create better socioeconomic conditions for its citizens.

As Africa's oldest republic, Liberia has very little to show for itself. It lacks the infrastructure that a country as old as it should have. This is because considering the country's age and wealth in natural resources, it should have been, for all practical purposes, the most developed country in Africa. But like many other African countries, corruption and the inequitable distribution of wealth, poverty, illiteracy, political instability, war, and other ills of society have held the country hostage from achieving economic prosperity.

Presently, the new government seems to be making strides in bringing some form of economic development to the country but is still struggling to curb illiteracy, poverty, and corruption. In fact one of the major accomplishments of the Sirleaf-led government was the gradual infusion of information and communications technology in some areas of government as well as hosting the country's first national ICT conference. This conference also led to the creation of a national ICT policy draft to serve as the country's guide as it shifts toward a new experience leveraged by electrons and information.

The next section provides a brief discussion of postwar Liberia. I have omitted several historical details that have no bearing on implementation or the lack thereof of ICT in Liberia. My intent is to shed light on issues that will kindle a debate on ICT and not the complete history of Liberia.

Postwar Liberia—The Poverty Reduction Strategy (PRS)

As peace and stability returned to Liberia, several postwar challenges have emerged. The issues of resettlement, rehabilitation, healthcare, education, national security, commerce, the environment, and other areas that are part of a country's recovery process present insurmountable challenges to the Liberian government as well as its people. Liberians have to tackle these challenges to ensure sustainable economic growth.

The election of a new president also brought about new plans to recover the economy. As part of her efforts to restore Liberia, President Ellen Johnson Sirleaf's government initiated the Poverty Reduction Strategy Program (PRS), which was developed by stakeholders of the country, including representatives from the rural sectors, the IMF, the World Bank, and the Liberian government. "A Poverty Reduction Strategy Paper (PRSP) is a national document that analyses the causes for poverty in a country and sets out a strategy to overcome them. A Poverty Reduction Strategy is meant to be a national process steered by the government and involving domestic stakeholders as well as external development partners" (Newborne 2004).

Since the election of President Ellen Johnson Sirleaf, Liberia has made great strides. According to an article in *The Economist*, the country's national budget increased from $80 million in 2006 to $350 million in June of 2010, and this has persuaded the International Monetary Fund (IMF) to erase the country's external debt of $4.9 billion. "Monrovia is witnessing a building boom, with beachside resorts and blocks of flats springing up, along with some conspicuously grand mansions belonging to well-known politicians. Ms. Sirleaf has also freed Liberia's forestry and diamond sectors from UN sanctions and renegotiated a controversial contract with a steel giant, ArcelorMittal," the article reports (*The Economist* 2010).

Some Data on Postwar Liberia

The table below illustrates some important geographic and economic data of postwar Liberia.

Geographic and Economic Data of Postwar Liberia

Indicator	Data
Areas	43,000 Sq miles
Population	3.9 Million
Population below poverty line	80% (2000 est)
Language spoken	English
Literacy	20%
Gross Domestic Product (GDP)	$1.627 Billion (2009 est.)
Real GDP growth rate	5% (2009 est)
Per capita GDP	$500 (2009 est)
Economic activity	Agriculture: 76%, Industry, 5.4%, Services, 17.7%(2002 est)

Monetary Unit	Liberian dollar
Main exports	Diamonds, iron ore, rubber, timber, coffee, cocoa
Exports	$1.197 billion (2006 est)
Export partners	India 26.5%, US 17.9%, Poland 13.9%, Germany 10.1%, Belgium 6.8% (2008)
Imports	$7.143 billion (2006 est.)
Imported commodities	fuels, chemicals, machinery, transportation equipment, manufactured goods; foodstuffs
Import partner	South Korea 27.2%, Singapore 25.5%, Japan 11.8%, China 11% (2008)
Labor Force	1.372 Million
Debt (external)	$3.2 billion (2005 est.)
Economic aid (recipient)	$236.2 million
Unemployment rate	85% (2003 est.)

Table 1.1—Source: *World Factbook* 2010

Liberia is also making strides in the area of infrastructural development. Several new businesses have opened, roads are being paved, schools being built and opened, exclusive markets are being established, and so on. Basic amenities, such as running water, electricity, and telecommunications, are being restored. Of all these utilities, telecommunication has improved dramatically and has been a major source of revenue. Nationally, telecommunications have achieved overwhelming progress in the last ten years. Cellular phones have become primarily the most ubiquitous means of communication in Liberia. Internet penetration, while very slow, is also making progress as many organizations and businesses have started to gradually install new satellite equipment for Internet access in Liberia.

Overall, relative progress has been made in Liberia despite the gradual pace at which some sectors have been impacted. For example, information and communications technology (ICT), which is also included in the country's Poverty Reduction Strategy document, has experienced little progress, although it is a very significant area. The creation of the Liberia Telecommunications Authority, the development of a national ICT policy document, and discussions about connecting to a submarine fiber-optic cable are but a few of the developments that have been made in the telecommunications and ICT sector of Liberia.

There is no doubt that ICT will bring economic development to Liberia, although this will only happen if there is shift in paradigm in all sectors of the country. In addition, the full support of Liberian stakeholders and the willingness of Liberians to project into the future without hesitation will also play a major role in transforming the telecommunications and ICT sector toward economic development in Liberia. Most of all, political will and leadership must be attached to this endeavor because they are crucial to the economic development. There is tremendous hope for Liberia as it embarks upon a recovery process that will lead to sustainable economic prosperity.

Chapter 2

Major Players In Liberia's Telecommunications And ICT Sectors: Yesterday And Today

This chapter is primarily based on pre- and postwar knowledge of Liberia's telecommunications and ICT sector. What follows is my perception/vantage point of telecommunications and ICT in Liberia influenced by personal knowledge and experience, and I must therefore submit that there is a great possibility that some information others may deem important may not have been addressed or referenced; for this, I apologize.

This chapter is intended to introduce the reader to the major players in Liberia's telecommunications and ICT sector since they will be referred to throughout this book. Previously, they were referred to as the telecommunications sector, but since mobile operators emerged providing ICT services, the sector is now referred to as the telecommunications and ICT sector. I will explore the impact the five major players have on Liberia's telecommunications and ICT sector. These five major players include the Ministry of Post and Telecommunications (MoPT), the Liberia Broadcasting System, the Liberia Telecommunications Authority (LTA), the Liberia Telecommunications Corporation (LIBTELCO), and mobile operators/Internet service providers. Of the five major players aforementioned, only the Ministry of Post and Telecommunications, the Liberia Telecommunications Corporation, and the Liberia Broadcasting

System existed prior to the civil war (Bernard 2004). The other two, Liberia Telecommunications Authority and the mobile/Internet service providers, emerged postwar. There will be limited discussion on the Liberia Broadcasting System; I choose rather to dwell on the other four players, since the context of the book relates to the nature of their operations.

Prewar Players In Liberia's Telecommunications And ICT Sector—MOPT, LBS, LTC

Prior to the information age, Liberia's involvement with what is now the global community was handled by the Ministry of Post and Telecommunications, the Liberia Telecommunications Corporation, the Liberia Broadcasting System, and several independent radio stations. The Liberia Broadcasting System was/is a government-owned agency that handled both radio and television broadcast services. In addition, there were other privately owned radio stations that provided services to the Liberian populace. In the area of telecommunications, the Liberia Telecommunications Corporation (LTC) served as the main "operator" for the country. It was (and still is) government owned and the sole agency responsible for the country's fi xed telephone and fax lines. It was established by the Legislature of Liberia in 1973 by the enactment of the Liberia Telecommunications Act. In 1978, that act was amended and led to the creation of the Ministry of Post and Telecommunications (Liberia 2007).

This ministry was given "carte blanche" to govern through policy and regulation over the telecommunication sector of Liberia. The Ministry of Post and Telecommunications also governed the postal services along with those indicated in the act of 1978. The creation of the Ministry of Post and Telecommunications left Liberia Telecommunications Corporation as a mere operator, which it still is today (Liberia 2007).

Current Liberian Telecommunications And ICT Sector

With two government agencies running the telecommunications sector, there were still several issues that needed to be addressed. The act creating both agencies did not illustrate a separation between the "policy maker," "the operator," and the "regulator." Other issues were the lack of standard licensing fees, provisions preventing monopolies, etc. Fortunately, in 2007, the national legislature of Liberia addressed these issues through the Liberian Telecommunication Act of 2007 and established the Ministry of Post and Telecommunications as the "policy maker," Liberia Telecommunications Authority as the "Independent Regulator," and Liberia Telecommunications Corporation as the "National Operator." The Liberia Telecommunications Corporation was also renamed LIBTELCO (Liberia 2007).

The Policy Maker (Ministry Of Post And Telecommunications)

As a founding member of the Universal Postal Union, the Ministry of Post and Telecommunications was known as the Department of Posts and Telegraphs in the early fifties. However, its name was changed to the Ministry of Post and Telecommunications through an act of the legislature in 1978 (Telecommunications 2009). The ministry, currently located in Monrovia, provides postal services to the Liberian people. Additionally, the Ministry of Post and Telecommunications is the arm of government that is responsible for developing policies for the country's telecom sector.

Like every government agency, the Ministry of Post and Telecommunications was severely impacted by the civil war. The ministry's headquarters was destroyed, looted, and served temporarily as a home to refugees during and after the civil war. The destruction caused by the war prevented the ministry from delivering postal services to the people of Liberia. Fortunately, the Universal Postal Union and other stakeholders including the government of Liberia and international organizations stepped in to revive the ministry. Hitherto the writing of this book, almost all counties in Liberia, according to the ministry's Web site, were being provided postal services. With new technologies being injected into the country, it is hoped that the Ministry of Post and Telecommunications also referred to as MoPT, will regain, if not surpass, its status as the hub of West Africa (Ministry of Post and Telecommunications 2009).

In the area of information and communications technology, within the Ministry of Post and Telecommunications is its Department of Telecommunications and Technical Services, which is charged with the responsibility of formulating policies and strategies for ICT and telecommunications. This arm of the ministry is expected to focus on ICT and telecommunication related to subregional, regional, continental, and global bodies such as the subregional West African Union known as ECOWAS, the African Union (AU), the International Telecommunication Union (ITU), WSIS, etc. The deputy minister for technical services heads this department and is assisted by the assistant minister for telecommunications, both of whom are appointed by the president based on the advice and consent of the Senate (Ministry of Post and Telecommunications 2009).

When I began writing this book, my brief conversations with a few employees at the Ministry of Post and Telecommunications and LIBTELCO seemed to focus only on connecting Liberia to the controversial underwater submarine cable (discussed later in chapter 3) and the impact it would have on Liberia. What was interesting during those conversations was the difference in opinion about how Liberia should connect to the SAT-3/WASC submarine cable, which was the cable of choice then. Some favored connecting through Cote d' Ivoire, while others preferred the Mano River Union, which is an amalgamation of three countries—Guinea, Sierra Leone, and Liberia.

However, the Ministry of Post and Telecommunications had explored the option of connecting to the SAT-3/WASC through the Mano River Union in collaboration with Guinea and Sierra Leone and with the help and support of the International Telecommunications Union. In addition to connecting to the SAT-3/WASC cable, the ministry plans to embark upon several initiatives that will allow broadband connectivity through a fiber-optic network in and around Monrovia and to the global community. This will grant greater access to broadband connectivity, significantly reducing the current prohibitive costs imposed by companies providing Internet service via satellite communications technologies (Ministry of Post and Telecommunications 2008).

With the advent of the Africa Coast to Europe (ACE) cable system, it seems that the ministry's plans to join the SAT-3 have changed. In March of 2010 the Ministry of Post and Telecommunications put out a release that presented its policy framework for fiber-optic cable in the country. This policy framework mandated that the Liberia Telecommunications Authority begin making plans for the ACE cable system (*TLCAfrica* 2010). The Africa Coast to Europe cable system as discussed in chapter 3 is expected to go live in 2012.

The Independent Regulator (Liberia Telecommunications Authority)

In 2004, during the interim government of Mr. Gyude Bryant, a new bill, "Bill No. 18," led to the temporary creation of the Liberia Telecommunications Authority (LTA) to act as the regulator of the telecom sector while the Ministry of Post and Telecommunications retained the function of policy maker (Best & Thakur 2008). The creation of the Liberia Telecommunications Authority was done in consultation with the World Bank.

In 2007, the Liberia Telecommunications Authority was made a permanent regulatory entity by the Liberian Telecommunications Act. Th e LTA is responsible for licensing all telecommunications operators and assigning civilian and noncivilian frequencies to applicants in consultation with the "policy maker," the Ministry of Post and Telecommunications (Liberia 2007). The LTA is also responsible for the implementation of a national numbering plan (a system used to assign telephone numbers) and ensuring fair and equitable competitive practices and is expected to regulate interconnection, co-locations, and general tariffs among other things (Liberia 2007).

As far as governance, the Liberia Telecommunications Authority is being run by five commissioners, with one of them serving as chairman. These commissioners are said to be well compensated, because, it is believed, they hold crucial positions that are vulnerable to corruption and that paying competitive salaries would prevent any inappropriate acts. Yet, governance has been a major problem since the agency's establishment. Up to the time when this book was conceived and written, there had been two failed administrations. The failure of those administrations was rumored to be related to corruption, which certainly makes the issue of paying lucrative salaries a paradox. Because it is seen as such a cash cow, it has become sought after by several Liberians as one of the most highly desirable entities of government for employment. I have always been concerned about this, because I believe that individuals seeking employment with the entity will be doing so only for the sake of personal gains and not necessarily the interests and future of the Liberian people. There have been several incidents in Liberian history where agencies of government considered to be "lucrative" ended up being officials' personal bank accounts instead of serving the purpose for which they were created. In fact, what is now LIBTELCO, the country's national operator, used to be a very lucrative institution but fell short of fulfilling its duties because revenue collected found its way into personal bank accounts instead of the appropriate coffers.

With two initial failed administrations, we can only hope that the current administration and future administrations will subsume the interests of the Liberian people and provide the services that they are hired to provide. Liberia cannot always rely on foreign institutions to manage its resources, because its citizens cannot manage those institutions without any discrepancies. LTA is a significant arm of government and is also crucial to the achievement of a digital Liberia.

The National Operator (Liberia Telecommunications Corporation)

I can vehemently argue that every part of Liberia was affected by the civil war. Similarly, I can also argue that there were some agencies of government that were already at the periphery of closure or destruction even before the civil war. The Liberia Telecommunications Corporation (LIBTELCO) is one of the many agencies of government that was gradually failing prior to the civil war. Faced with management challenges and other issues that led to its dormancy, including lack of fiscal discipline, the corporation could not handle the infrastructure or the functions for which it was formed (LIBTELCO 2009). This led to a deterioration of the telecom infrastructure, which was further exacerbated by the looting and vandalism that characterized the civil war. LIBTELCO remained nonoperational until the ushering in of the administration of President Ellen Johnson Sirleaf. The new government took immediate action by appointing an interim board of directors to assess the technical and financial viability of the corporation and work out modalities that would lead to the restructuring/revamping of the telecommunications sector in Liberia (LIBTELCO 2009). This was the impetus for the development of the Telecommunications Act of 2007 and the appointment of a permanent board and management team.

LIBTELCO being the national operator is expected to provide fixed wireless phone, wireless Internet, and fax services to the public. The company is expected to be the provider of services to the government of Liberia as well as to compete with other telecom firms for the nongovernmental market. LIBTELCO has made some strides toward that end since the postwar management team was put in place. Sometime in 2009, it announced the deployment of a state-of-the-art CDMA2000 1X-EVDO network and commercial services (LIBTELCO 2009). CDMA, which stands for carrier division multiple access, is technology that was first proposed by Qualcomm and is being used as the standard for cellular service in North America. Compared to global system for mobile communications (GSM), CDMA provides better signals and secure communications. CDMA2000 1X-EVDO (evolution-data optimized) is a CDMA-based technology created recently by Qualcomm. This new technology provides higher packet transmission rates and is used widely by wireless operators. CDMA2000 1xEV-DO was approved at the International Telecommunications Union (ITU) Stockholm Conference in 2001 as an IMT-2000 standard according to the CDMA Development Group (cdg.org n.d.).

Despite these improvements, LIBTELCO faces several challenges to ensure its viability and relevance. Also, prior to the war there had been discussions regarding the privatization of LIBTELCO. While many argue that privatizing LIBTELCO would be a better alternative, I have heard several Liberian authorities on several occasions indicate that there is no need to privatize LIBTELCO considering its ability to sustain itself. Hitherto the writing of this book, nothing had happened to that effect. But whatever decision is made regarding this agency, it is my hope, and I believe that of every Liberian, that the result will favor Liberians ... after all, it belongs to them.

Mobile Operators and Internet Services Providers

A few years after Liberia's general elections and realizing relative peace and stability, Liberia has become the quintessential wireless telecommunications market where communication is mostly done using cellular phones. The country currently has four major mobile operators that compete for customers. Internet service is available through various Internet service providers, and the mobile networks use GPRS, EDGE, and WiMAX technologies. The country lacks basic infrastructure; hence satellite communications technologies are being used widely, making the cost of connectivity extremely prohibitive and a barrier to ICT penetration. Despite this, mobile technologies have gained higher penetration in Liberian markets than information technologies. The good news is that Liberia is expected to benefit from the arrival of a new international submarine fiber-optic cable through the Africa Coast to Europe (ACE) project in 2012. The ACE project is discussed in chapter 4.

The introduction of mobile technologies in Liberia brought an unprecedented and significant change to the Liberian society, because it bridged the communications gap that had existed for years between rural and urban areas in Liberia and between Liberia and the global community. The cellular phone, which is the most common form of communication in Liberia, credits its advent to the Charles Taylor administration. It is during this time that the private mobile operator known as the Lone Star Communications Corporation implemented its GSM World in the country to initiate what would be a communications revolution. Since then, several mobile operators and Internet service providers have entered the Liberian market. Those who came later included Cellcom, Comium, and Libercell. These four major mobile operators were given privileges to utilize the telecommunication spectrum in Liberia prior to the election of in 2005. Of course, at the time, the entire spectrum was not properly regulated, and these mobile operators were allowed to set up their towers for operations without any policy in place.

Lone Star, a GSM (global systems for mobile communications) company that began operating in Liberia in 2000, is owned by a South Africa company known as MTN (Ministry of Post and Telecommunications 2009). Cellcom, another GSM company, is owned by a partnership of American and Israeli investors (Ministry of Post and Telecommunications 2009). Comium, yet another GSM company, is owned by Dalloul Group of Lebanon, while Libercell, also a GSM company, is owned by Lebanese investors (Ministry of Post and Telecommunications 2009). I refer to these companies as GSM companies because they use the GSM or Global System for Mobile communication technology, which was developed in France in 1980, the year Liberia experienced its military coup. GSM, unlike CDMA, is a 2G system based on FD-TDMA, or frequency division–time division multiple-access radio access. It is used widely for wireless cellular communications but lacks the capability to seamlessly work in data networks, thus making it not so ideal for Internet services.

Just as other businesses of the Liberian market are dominated by non-Liberians, so too are the mobile companies dominated by non-Liberians. It is no surprise that Lebanese investors own a considerable amount of the mobile market because historically they have been major contributors to the Liberian business sector. But as the telecom sector grows, Liberians will begin investing in that area and will be even more confident when the country links to a submarine fiber-optic cable system for broadband connectivity. As I mentioned earlier, this will be done through a connection to any one of the submarine fiber-optic cables located along West Africa, most likely the ACE system.

Cellular phones have become an essential part of daily life in Liberia. Rural Liberia, which has a high percentage of the country's population, now benefits from mobile communications because of the convenience it brings and the new capacity residents of that part of Liberia have gained. Cellular phones have brought a paradigm shift that has changed the Liberian way of living.

Furthermore, the advent of mobile technologies in Liberia has also had a significant impact on the country's economy. New jobs were created when private mobile operators arrived, and a considerable number of Liberians earn a daily living by selling cellular phones, scratch cards, and other mobile phone accessories. Businesses in Liberia have also benefited from this form of communication; their organizational culture and processes have experienced a major change.

Each mobile operator has its own group of users. In the United States, you pick a provider based on certain features available on a plan. In Liberia, it is the type of network provider or type of phone you buy that aligns you with a particular mobile operator.

Liberians are currently enjoying services that parallel those in Western or developed countries because of the improvements that have been made in the telecommunications and ICT sector. Roaming services, SMS, MMS, and Internet access, etc., are being provided by mobile operators and Internet services providers.

While the implementation of broadband connectivity is being explored for lower costs, convenience, coverage, and speed, cellular phones (through satellite communications) still dominate the market. This is because they require less infrastructure and afford more flexibility and portability. Also, the Liberian mobile market is fully liberalized, giving room to a plethora of companies to engage in the sale of telecommunications and ICT products and services. The liberalization of the telecom and ICT market necessitates stronger and more aggressive regulatory engagement with existing and potential firms. This is where the Liberia Telecommunication Authority (LTA) will play a major role. In addition, LTA will also have to deal with the challenges that were inherited at its formation. Some of these challenges include addressing policies set forth initially that were either too weak or did not properly address the use of telecommunications spectrum in Liberia by firms. In some cases, there might be a need for newer polices and regulations to govern a new and reformed telecommunications and ICT sector. Redeeming the fair value of licenses that were given to companies prior to the establishment of LTA would pose a major challenge. Also, there are several other aspects of the mobile market that still need to be fully regulated and that make the job of LTA a humongous and a challenging one.

There are, and have been, several initiatives taken on by various Liberians to improve and enhance telecommunications and ICT penetration in Liberia. Some of these initiatives include the opening of local computer schools, Internet cafés, and so on. International scholarly and research initiatives have also been part of this process; for example, several research work done by Dr. Michael Best and a number of doctoral students at Georgia Tech have been published in the *Association for Computing Machinery (ACM)*. Also, in 2007 Liberia held its first ICT stakeholders conference in Monrovia, which brought together Liberian ICT professionals as well as companies such as Cisco Systems, Microsoft, etc. Following that meeting the National ICT and Telecommunication Draft was formulated. This draft, which was also made public for input from Liberians both locally and internationally, is referenced in chapter 16, and I was honored and privileged to have made some contributions as was requested by the Ministry of Postal and Telecommunications.

A lot has happened since the election of President Ellen Johnson Sirleaf. As the country moves ahead a lot of new technologies will be available, especially with the advent of broadband access through connection to a submarine fiber-optic cable system and a galvanized telecommunications and ICT sector.

Chapter 3

The Submarine Cables of West Africa: Conduit to Liberia's Digital Revolution

The advent of the Internet in Liberia kindled a new scenario; a scenario that necessitates Liberia's entry into the digital community for economic growth. Entry into the digital community for economic growth had been barricaded by the proverbial "digital divide." In bridging the digital divide, Liberia gravitated toward the only logical option available—the use of satellite communications to gain access to the outside world. While satellite technologies are very convenient and logical for countries without infrastructural capacities, they are expensive and have their own share of challenges.

Cost is a major factor and barrier to ICT penetration in developing countries. In order to develop economically in the twenty-first century, access to the digital community is imperative, thus necessitating the exploration of a cheaper and better-quality medium/media of connectivity than the expensive satellite communications technologies that Liberia is currently utilizing. This alternative medium is a submarine fiber-optic cable situated along the coast of West Africa.

In 2009, after reading the ICT Development Index report, which is a periodical that does a comparison on developments in information and communication technologies (ICT) in 150 countries over a five-year period from 2002 to 2007 (Union 2009), I noticed that the

International Telecommunications Union (ITU) referenced several West African countries, including Gambia. There was no mention of Liberia, although personally I believe that during that period the country had made significant progress in the area of telecommunications and ICT. I presumed that the relative progress that was made in Liberia with regard to ICT penetration was not enough to earn it a space in the ITU's report. But current efforts to link the country to a submarine fiber-optic cable for broadband connectivity raise hopes for me and every Liberian. And I believe that a better connection to the global community is imminent and that economic development in a digitally bridged Liberia is soon to come. Broadband connectivity is essential to Liberia's economic development in the information age. It is for this reason that connecting to a submarine fiber-optic cable is crucial and what I see as the conduit to Liberia's digital revolution.

This chapter briefly discusses submarine fiber-optic cables that are situated around the coast of West Africa and are "within proximity" of Liberia. I will spend more time on the South Atlantic Terrestrial Cable 3/West Africa Submarine Cable (SAT-3/WASC) and the Africa Coast to Europe (ACE) project because these cables regularly spark controversy during Liberian roundtable discussions. I also briefly discuss a few other existing and planned submarine cables in the West African region, including ATLANTIS-2, GLO-1, Main One, and WACS, as listed by Wikipedia in 2009. When this book was being written, TeleGeography—a submarine fiber-optic research firm—reported that there are 111 submarine cables in the world and 19 others are being planned for 2011 (TeleGeography 2010). Submarine cables in Africa were included in those numbers as well.

SAT-3/WASC/SAFE

The South Atlantic Cable 3/West Africa Submarine Cable/ South Africa Far East cable (SAT-3/WASC/SAFE) is a submarine communications cable that links from Portugal and Spain to South Africa and then connects to several West African countries along the route (SAT-3/WASC/SAFE 2009). In April of 2001 when it began operations it provided the original link between Europe and West Africa. I was abroad at the time of the installation and later discovered from friends and family in Liberia that Liberia was not able to connect to this submarine fiber-optic cable. I learned that the reason for this missed opportunity had much to do with the civil war and the colossal cost of installation.

According to information on SAT-3/WASC/SAFE's Web site, the SAT-3/WASC/SAFE cable is a combination of two subsystems, which include the SAT-3/WASC and the South Africa Far East cable (SAFE), located respectively off the west and east African coasts. SAFE is also a submarine fiber-optic communications cable that links South Africa to Asia. SAT-3/WASC is situated in the Atlantic Ocean, and SAFE is situated in the Indian Ocean. The combined length of both segments of SAT3/WASC/SAFE is 288,000 km (SAT-3/WASC/SAFE 2009).

SAT-3/WASC/SAFE has cable landing points in Portugal, Spain, Cote d' Ivoire, Ghana, Nigeria, Cameroon, Gabon, Angola Senegal, Benin, a Reunion, Mauritius, India, Malaysia, and South Africa (SAT-3/WASC/SAFE 2009).

In 1993, the SAT-2, which was the first sub-Saharan Africa submarine fiber-optic cable system installed, bypassed the entire west coast of Africa. Its successor, the SAT-3, made it along the west coast of Africa with landing points installed in several West African countries excluding Liberia (Ruddy 2007). This missed opportunity is what holds Liberia's digital revolution hostage.

While the SAT-3/WASC was received as a significant breakthrough in African telecommunications, it has not been able to bring marked improvement to the Sub-Saharan region as was previously anticipated (Ruddy 2007). African countries that are landlocked have to connect to cable landing points of other countries that are within proximity of the SAT-3/WASC. The problem with this is that the incumbent countries that own the cable landing points have a monopoly that allows them to charge heavy fees. As a result of this high cost—which has been the barrier to ICT penetration in this region—many countries faced with this problem have resolved to use satellite technologies, which have limitations, and it is this situation that has led to the advent of additional submarine cables (Ruddy 2007).

In Liberia conversations about connection to a fiber-optic cable had always gravitated toward the SAT-3/WASC. While a connection to the SAT-3/WASC cable will connect Liberia to the world, there are several complexities involved in choosing this option. SAT-3/WASC is owned by a consortium of stakeholders yet to liberalize the use of the cable system that will allow a flexible distribution of bandwidth in countries that have access to the cable (Ruddy 2007). Liberalization will also kindle the advent of companies that will provide more alternatives for citizens of their respective countries.

The SAT-3/WASC/SAFE consortium consists of thirty-six stakeholders, and installation is said to have costs of about $600 million. The consortium of shareholders will own the cable for the next twenty-five years, according to Fiber for Africa (Fiber for Africa 2009). The SAT-3 Consortium has three operational subcommittees: finance and commercial (chaired by Ghana Telecom); an operations and maintenance subcommittee, and a delivery and restoration committee, both of which are chaired by France Telecom (Southwood 2006). s The largest investors in the SAT-3/WASC/SAFE cable system: France Telecom (12.08 percent); Nitel (8.39 percent); TCI, a subsidiary of AT&T (12.42 percent); and 8.93 percent for VSNL (Southwood 2006).

Out of the thirty-six members of the SAT-3 consortium as reported by Fiber for Africa, Africa also has twelve investors in this consortium. They include Senegal (Sonatel), Côte d'Ivoire (Côte d'Ivoire Telecom), Ghana (Ghana Telecom), Benin (OPT), Nigeria (Nitel), Cameroon (Camtel), Gabon (Gabon Telecom), Angola (Angola Telecom), South Africa (Telkom), La Reunion (France Telecom), Mauritius (Mauritius Telecom), and Namibia (Telecom Namibia). The managing agent for the African section of the cable is South Africa's Telkom. The consortium also controls the price charged for international bandwidth, and these prices vary from country to country (Fiber for Africa 2009). I believe that prices are charged per mbps (megabits per second).

In recent years after the first missed opportunity, a connection to the SAT-3/WASC had been explored by the Liberian government; obviously nothing happened and there was no connection. While a connection to the SAT-3/WASC would provide broadband connectivity to Liberia, there may be other issues that might not be favorable to the Liberian government and other stakeholders. I will not touch on that in this book, although from a technical standpoint, the alternatives to SAT-3/WASC—the ACE System and the WACS—provide a new technology and higher capacity, thus making them better options for Liberia.

THE ACE SUBMARINE CABLE PROJECT

The ACE submarine communications cable for now (at the time this book was written) is a planned cable system along the west coast of Africa with connections between France and South Africa (Wikipedia, ACE [cable system] 2009). According to a press release from Orange—a major stakeholder in the project—dated December 1, 2009, countries that are to benefit from ACE include France, Spain, Portugal, Morocco, Canary Islands (Spain), Western Sahara, Mauritania, Senegal, Gambia, Guinea-Bissau, Guinea, Sierra Leone, Liberia, Côte d'Ivoire, Ghana, Togo, Benin, Nigeria, Cameroon, Sao Tome and Principe, Equatorial Guinea, Gabon, Congo, Angola, Namibia, and South Africa. The press release also mentioned that the submarine cable system is expected to be more than 14,000 km long and will be operational in 2011 with a minimum capacity of 1.92 Tbit/s (Orange 2009. The previous RFS (ready for service) date of 2011 would later be changed to Q2 of 2012.

ACE comes with wavelength-division multiplexing (DWDM), which works seamlessly with existing submarine cables. The same press release goes on to list the twenty-five telecommunications operators that are parties to the ACE consortium (Orange 2009).

Compared to the SAT-3/WASC/SAFE mentioned above, ACE provides more benefits for countries that are expected to gain connection to it. For countries like Liberia, Guinea, and Sierra Leone, which do not currently have any cable landing or access to a submarine cable, the ACE cable system will provide direct access to broadband connectivity. ACE's impact on the economies of these countries will be significant, because access to broadband by the countries connected will serve as a catalyst to their digital revolution.

Countries that are using the SAT3/WASC/SAFE submarine cable will have the ACE cable system as an alternative route, which will give them a higher level of redundancy in the event the SAT-3/WASC/SAFE fails, as it did earlier in 2009. It can be recalled that in July of 2009 the SAT-3/WASC/SAFE cable system experienced an outage due to a damage done to the cable system that affected services in Togo, Niger, Nigeria, and Benin (Heacock 2009).

Upon completion of the ACE cable system, cable landing points will be in the following countries: Monrovia, Liberia, Cape Town, South Africa, Penmarc'h, France, Lisboa, Portugal, Asilah, Morocco,

Tenerife, Canary Islands (Spain), Nouakchott, Mauritania, Dakar, Senegal, Banjul, Gambia, Sucujaque, Guinea-Bissau, Conakry, Guinea, Freetown, Sierra Abidjan, Leone, Côte d'Ivoire, Accra, Ghana, Lomé, Togo, Cotonou, Benin, Lagos, Nigeria, Kribi, Cameroon, Santana, Sao Tome and Principe, Bata, Equatorial Guinea, Libreville, Gabon, Muanda, Democratic Republic of Congo, Luanda, Angola, and Swakopmund, Namibia (Wikipedia, ACE [cable system] 2009). Hitherto the writing of this book, the ACE system remained Liberia's hope of gaining broadband connectivity.

As I mentioned above, in early 2000, Liberia missed an opportunity to gain access to broadband connectivity via the submarine fiber cable, SAT3/WASC. Fortunately, Liberia has been presented with another opportunity; the ACE submarine fiber cable system. The dreadfully slow access to the Internet in Liberia today is a direct result of that missed opportunity. As I write this book, Internet bandwidth has been and is still a major problem in Liberia. The lack of political will, capacity, and infrastructure has deprived the country of broadband services, which are needed for economic growth. Currently, communication is done via satellite, which has prohibitive costs attached to it. In Florida, where I previously resided, I had a one MB (upstream) connection to and a two MB (downstream) from the Internet for the cost of a little over $20.00 per month. If I were to have the convenience of a 2 MB Internet connection in Liberia at this time, I would probably be paying several thousand dollars.

Previous discussions among some of my colleagues regarding various options to connect Liberia to a submarine fiber-optic cable included a connection to the SAT-3/WASC cable landing point in Cote d' Ivoire, a connection to AT& T's cable landing point in Senegal, a direct connection to the SAT3/WASC, and so on. These options have never transcended mere formal or informal discussions, of course. The ACE cable system seems to bring hope to Liberians for broadband services, and it appears that the Liberian government will gravitate toward this option as it provides a better option to bringing broadband to Liberia.

ATLANTIS

The ALANTIS submarine fiber-optic cable is a fiber-optic transatlantic telephone cable connecting several countries, including Argentina, Brazil,

Senegal, Cape Verde, Canary Islands, and Portugal. This initiative was organized by Embratel and was activated in 1999. It is about 13,100 kilometers in length. ALANTIS has a capacity of 19,840 Mbits/s and has landing points in Argentina, Brazil, Cape Verde, Senegal, Spain (Canary Islands), and Portugal. The official network administrator is Telecom Argentina (Wikipedia, ACE [cable system] 2009).

GLO-1

Globacom-1 submarine fiber-optic cable, a.k.a. GLO-1 or Globalcom-1, is yet another submarine cable that is installed along the coast of West Africa between Nigeria and the United Kingdom. It is 9,800 km long and became operational in 2009 with a minimum capacity of 640 Gbit/s and can be expanded to 2.5 Tbs/s (Nwankwo 2009). Currently, landing points from the Glo-1 can be found in the following countries: Nigeria, Ghana, Senegal, Mauritania, Morocco, Portugal, Spain, and the United Kingdom.

The implementation of the GLO-1 makes Globalcom the first telecommunication company in the world to own its submarine cable without being part of a consortium (Nkanga 2009). Normally, submarine cables have always been owned by a consortium of Telcos or stakeholders, but Globalcom seemed to have changed that practice. Its presence in West African countries, especially Nigeria, is expected to lead to a significant transformation in the areas of health, commerce, and governance because of the cheap bandwidth that it will provide (Nkanga 2009).

Main One Cable

Main One submarine cable is an submarine cable system that stretches from Portugal to West Africa with landing points along the route in various West African countries including Ghana and Nigeria (Main One 2010). The first phase, led by the Nigerian Main One Cable Company, is expected to be launched July 1, 2010, and spans 6,800 km from Portugal to Nigeria and Ghana. The cost of the project, according to Main One, is about $400 million USD. "The Main One Cable Company is wholly owned by African investors—African Finance Corporation, Nigeria; Pan African Infrastructure Development Fund, South Africa; FBN

Capital, Nigeria; Skye Bank, Nigeria and Main Street Technologies, Nigeria, which is the project sponsor" (Main One 2010).

The Main One project has the ambitious goal of not only delivering high-speed bandwidth of 1.93 Tbts/s but also providing several pioneering commercial opportunities for connecting countries and countries aiming to connect to it. According to the company, this includes providing access to customers in the West African region at prices less than 50 percent of current wholesale capacity prices.

Main One is to carry out this project in phases. Phase 1 of the project is expected to join Portugal to Nigeria and include the installation of a landing point in Ghana. It is estimated that this phase would cost $240 million USD. Expansion of Main One will be done subsequent to phase 1 and will connect South Africa, Angola, Gabon, Senegal, DRC, Cote D'Ivoire, and Morocco (Main One 2010).

WACS

According to Reuters (2009) the West African Cable System submarine cable is also expected to be ready for service (RFS) in 2011. The cable system is a $600 million submarine project that is expected to further boost international bandwidth capacity in Africa. It is expected to be 14,000 km long and carry a capacity of 3.84 Tbts/s (Reuters 2009). Angola Telecom, Broadband Infraco, Cable & Wireless, MTN, Telecom Namibia, Tata Communications (Neotel), Portugal Telecom, Sotelco, Togo Telecom, Telkom SA, and the Vodacom group constitute the consortium that owns WACS (Balancing Act 2009).WACS will land in the following countries: South Africa, Namibia, Angola, the Democratic Republic of Congo, (DRC), Canary Islands, Cameroon, Nigeria, Togo, Ghana, Cote d'Ivoire, Cape Verde, Portugal, and the UK. WACS is expected to be better than SAT-3/WASC and will cut down the high cost of Internet tariffs (Balancing Act 2009).

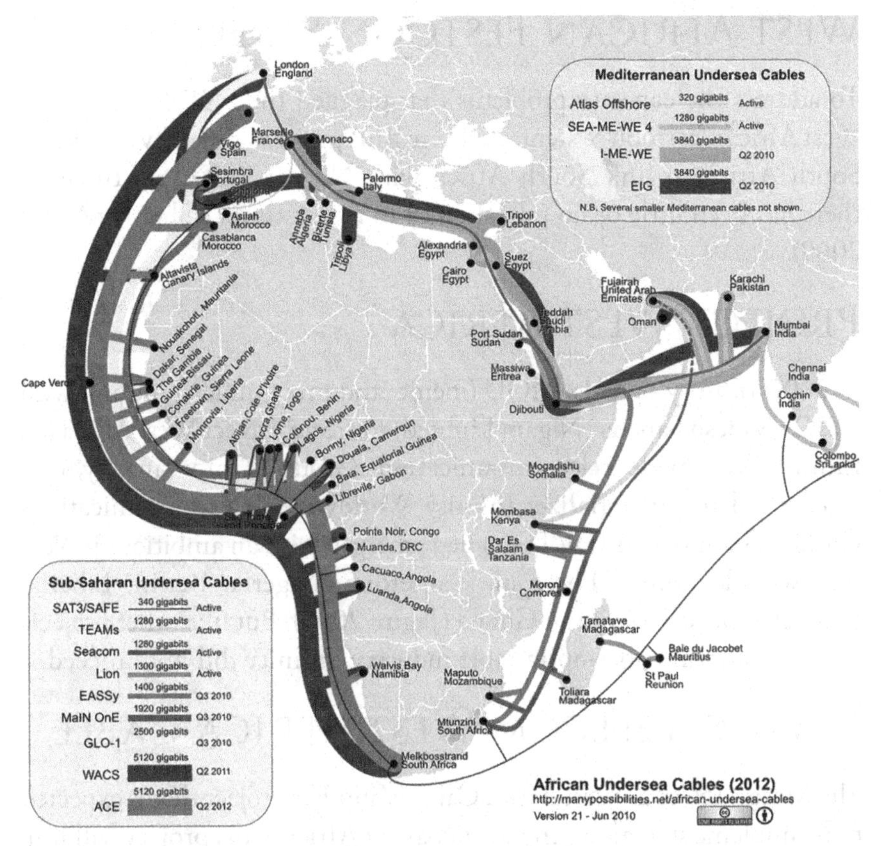

Figure 3.1. African Submarine Cables (2011)—Courtesy of Steve Song

I later discovered that WACS became an amalgamation of the following previously planned projects: Infraco AWCC, SAT4, and UhuruNet.

Other Submarine Fiber-optic Cable Projects For West Africa

Below are several submarine fiber-optic cable systems that were planned but have not been implemented. Some of these projects combined with other projects to become larger cable systems such as WACS and ACE.

West African Festoon System

To address the capacity problems that plagued the SAT-3 system, the West African Festoon System was born. This project was led by a Telkom-South Africa to link South Africa with Nigeria, but unfortunately, there has been little or no talk about it in recent times (Fiber for Africa 2009).

Project West Africa

In 2007, Infinity signed a MOU (memorandum of understanding) with VSNL (Videsh Sanchar Nigam Limited) to create Project West Africa (an Infinity Worldwide Telecommunications Group). Unfortunately, that project did not materialize. Infinity Worldwide Telecommunications Group of Companies (IWTGC) had emerged with an ambitious project that would connect Portugal, Cameroon, Nigeria, Benin, Liberia, Senegal, Cote d'Ivoire, and Ghana (Jagun 2009). But like other projects in the submarine fiber-optic cable industry, Infinity did not proceed.

Maroc Telecom West Africa Cable

The Maroc Telecom West Africa Cable is another project that is expected to be implemented along the west coast of Africa. This project will link several countries in West Africa with Morocco. These countries include Mali, Gabon, Burkina Faso, and Mauritania (Dow Jones 2010). Major shareholders in this project include the French media and telecom conglomerate Vivendi and the government of Morocco (Dow Jones 2010).

Uhurunet

As I mentioned above, the Uhurunet project merged with the WACS project.

SAT-4

The proposed SAT-4 project, as mentioned above, transformed with other projects into the WACS project (Jagun 2009). Rumor had it that the SAT-3 cable was about to run out of capacity hence, the need for an upgrade. There was also news that Telkom, which operates the SAT-3/WASC/SAFE submarine cable, had explored and realized the necessity of building a terabit- per-second "SAT-4" cable (Creamer 2007). Telkom had already upgraded from 40 Gbit/s to 120 Gbit/s.

There have been many discussions regarding other submarine cables in Africa, many of which did not materialize or ended up being harmonized with other projects. Overall, the barrage of submarine fiber-optic cables that have been erected, are being erected, or are planned in West Africa clearly indicates that there is business potential in the region.

Cable Landing In Liberia

What is a cable landing? A cable landing is the point at which a submarine cable makes a landfall. Landing points are special areas because they have to protect the cables from destruction by passing vessels or from natural causes. Liberia does not have a cable landing even now as I write this book. But in a March 2010 press release posted on the TLCAfrica Web site, the Ministry of Post and Telecommunications instructed the Liberia Telecommunications Authority (LTA) to create a national consortium that would link with the ACE consortium while it locates the "appropriate maintenance authority for the Liberia landing station" (TLCAfrica 2010). This demonstrates the Liberian government's efforts to connect to the ACE cable system.

Other Issues Regarding Broadband In Liberia

Because there is no cable landing point from the SAT-3/WASC for reasons I previously mentioned, there are several alternatives that are available at this point. As mentioned above, ACE will be putting a drop in Liberia; but this is expected to happen in 2012. That development will positively impact and bring major change in the West African economy.

There are also neighboring countries that have cable landing from where Liberia can link. The Cote D'Ivoire has a landing point that could be connected to a fiber backbone in Liberia, creating several access points for ISPs and other institutions. This will ensure that there is equitable distribution of bandwidth.

Another alternative is connecting through Senegal to the ATLANTIS, which LIBTELCO had mentioned on its Web site. I am not sure how cost-effective this would be, but I would assume that Liberia might want to partner with other countries to share the cost of this initiative.

Whatever solutions Liberia chooses there will be a need to keep some of the existing satellites infrastructure for redundancy. Redundancy will be needed with respect to international connectivity, since there is a possibility that the submarine cable system might experience downtime. This will prevent a situation in which there is a breakdown in the fiber link as happened in July 2009 with the SAT-3.

There will also be a need for an Internet exchange point that will serve local Internet traffic in Liberia. This will allow local Internet service providers and network operator(s) to easily exchange traffic within Liberia instead of having to go through international exchanges to route local traffic.

Strong policies and regulations on how access to the cable is determined will be needed whenever Liberia connects to the ACE system. The Liberia Telecommunications Authority being the regulatory body should also provide opportunities that will allow competitive pricing.

Although Liberia does not boast of a list of assets, it seems to be achieving a stable democracy. The illiteracy rate remains high, although significant work is being done to reduce it. The penetration of ICT into rural areas will greatly help in reducing this high rate of illiteracy.

Infrastructure and capacity are issues will need to be addressed. And technological advancements will continue to provide solutions to complex modern challenges that are seemingly being exacerbated by the application of traditional solutions.

With broadband connectivity and a fully liberalized market, the markets for Internet and international services will experience a boom. Government should encourage privatization of the sector in order to allow reforms that will lead to economic growth. Full liberalization of the telecommunications markets and allowing open access to a fiber-optic link to potential operators will allow the country to fully utilize the capacity allotted to the country.

Government regulators will have to ensure that capacity allotted to Liberia is fully utilized and not wasted as is being done in some other West African countries. For example, Cameroon's Camtel is said to be using about 50 percent of Cameroon's allocated capacity. This amount is said to be about 80 percent or more of the capacity used in the country, excluding what is used by other companies that are directly connected to the SAT-3 (Jagun 2008).

Submarine fiber-optic cables carry more than 80 percent of all overseas voice, fax, and Internet communications. Moreover, they offer better efficiency, reliability, and security than satellites. They (submarine cables) dramatically reduce the cost of bandwidth measured in dollars per megabit per second. They also reduce the round trip times (RTT) significantly by using shorter-distance terrestrial routes and reduce congestion because of their high capacity.

Currently many Internet users in Liberia access the Internet through independent satellite connections (VSATs), and this creates a serious problem because local traffic is routed through Europe and service is oftentimes slow. Hopefully, with the presence of more broadband connectivity, this issue will be addressed and more businesses/organizations/ISPs will ultimately migrate to fixed broadband.

For broadband connectivity to reach Liberia, there will be a need for all stakeholders to contribute to the process: UN, ADB, USAID, IMF, World Bank, local operators, local financial institutions, and so forth. While writing this book, I learned that there are negotiations being made for Liberia to join the Africa Coast to Europe (ACE) project for connectivity in 2012. The government has been less vocal about the ACE project, making Liberians skeptical of the possibility of this happening. This skepticism is a result of Liberia not making good on promises made and discussions had regarding connectivity to a submarine fiber-optic cable. But I have been optimistic about the ACE project; I believe Liberia will not miss this opportunity after missing the opportunity to connect to the SAT-3/WASC.

From now on, most of the chapters in this book will basically provide ICT solutions that are based on the implementation of a submarine fiber-optic initiative in Liberia. I stress the need for a submarine fiber-optic cable because it is indeed the conduit to a digital Liberia.

Part II
Education

Chapter 4

How Information And Communications Technology (ICT) Will Impact Education In Liberia

In 2006, as a visiting lecturer of the University of Liberia in a Management 305 class (Principles of Information Systems at the time), I predicted that the road to technological advancements in Liberia would be promising and that Western technology will penetrate Liberia sooner than we had anticipated. I also remember informing the students that the road will be less rocky if those in charge think 70 percent for Liberia and 30 percent for themselves instead of 30 percent for Liberia and 70 percent for themselves. What I meant then was that those in charge put Liberia first and themselves second, because ultimately, if they look at the big picture, everyone benefits when Liberia succeeds.

Several years have passed since that lecture, and as I predicted, Liberia has made some strides in terms of integrating technology into various sectors of the society. I am still of the opinion that Liberia will achieve modernity through technological advancements, the kind that we never could imagine twenty years ago. However, this will only be achieved if several conditions are met, one of which I mentioned above. One area of technology where its infusion will prove beneficial for the country and its society is in the area of education. Liberia obviously has a bright future in the "new economy," despite the views of cynics and those who oppose change. In this chapter, I will take you into the future of education in Liberia and how it will be impacted by technology.

Prior to the civil war (and currently), schools in Liberia relied on traditional instructional media: blackboard, chalk, and teacher-driven lessons. Computers and the Internet were unheard of. In fact, the typewriter existed as the main form of technology used to prepare tests, report cards, and instructional/printed material. Even private schools did not have the technologies that some of them currently have today. Because Liberia did not have access to up-to-date materials, I believe, the quality of education that the public/private school students received was subpar. Students relied heavily on their instructors, who in most cases were not college graduates or qualified to be in the classroom. Instructional materials provided by those teachers could not be contested or critically analyzed by students because they (students) did not have access to supplemental materials that could buttress their argument. But all of this will change! The advent of the Internet and advancements in technology and telecommunications will bring a new approach to education in Liberia. And this approach has already begun and being implemented through modern technologies.

The changes that are being brought through technology in the Liberian educational system are not only being implemented by the government or local stakeholders. Instead, they are being pioneered by schools' alumni associations that are based in the Diaspora. The B. W. Harris School Smart Technology Project which is discussed in chapter 8 is a quintessential example of how Liberians have decided not to rely solely on government for educational development. Currently, not all the schools in Liberia have integrated technology into their curricula; only a few have. Even the higher institutions of learning have yet to implement new instructional approaches that parallel those in the global setting. This lack of technology integration is a direct result of funding, infrastructure, and capacity. Personally, from my assessment, funding may not be the main problem, since several NGOs are willing to embark upon projects that can bring economic development. With funding, infrastructure can be built. So I believe that the lack of will and capacity are the main problems that have stalled the process of change in the educational system. My experience at B. W. Harris School is proof that the lack of will and capacity are preventing twenty-first-century approaches from reaching schools in Liberia. But all of this will change as we continue to engage, address, and predict challenges of the information age.

Despite the challenges Liberians are faced with, there is the potential for the country to shift toward twenty-first-century educational paradigms. In a few years, developments that are slated to occur in the area of ICT will immensely impact education in Liberia. These changes will allow Liberian schools to migrate toward twenty-first-century pedagogy where teachers serve as facilitators, thereby fostering an environment of student-driven learning and critical-thinking skills. Students will work in groups or teams, utilizing technology to connect, communicate, and collaborate with other students around the world in real time. This approach will improve education in Liberia as well as provide several benefits for the entire country. Now I am not saying that there should be an effort to use technology to replace educators but rather that we should use technology to enhance educators' instructional approaches in the classroom.

Currently, Liberia's workforce is more advanced in the area of technology than is the current school system. Students on the other hand have access to more modern technology than what is available in schools. Therefore, when they go to school, they become subjects of a learning environment whose state-of-the-art instructional media are still the blackboard and the chalk. But this will and must change! The decreasing cost of hardware/software, green technology, open source technology, cloud computing, and other new technologies will facilitate this move toward digital learning.

Already, Liberia has made significant improvements in the area of telecommunications as evidenced by the limited availability of Internet service and cable television in some parts of the country, especially in the capital city of Monrovia. Moreover, connection to a submarine fiber-optic cable situated around West Africa will lead to the injection of triple-play services (voice, data, and video) in the Liberian educational sector, allowing us to transcend traditional learning approaches and arrive at more modernized and digitized ones.

Technology will bring the changes that will give Liberia competitive advantage over other countries, or at least level the playing field—if not globally, then regionally. Schools, through the help of all stakeholders, will help to provide the type of education needed to prepare Liberians for the digital economy.

As Liberia moves forward, the educational environment will change immensely. There will be an infusion of technology that will include the use of interactive software, Web 2.0 technologies, and other technologies that will enhance learning. The Internet will be crucial in the Liberian classroom as well. With broadband Internet access, schools will take full advantage of the resources available on the Internet. For example, there is a growing number of online Liberian radio stations as a result of the advancement in and the availability of modern technology. With broadband, institutions of learning can use educational broadcasting stations to deliver instruction to students, including kindergarten students, who learn better through audio and visual materials. College students will learn thermodynamics and econometrics through videos or lectures from professors at universities around the world through the Internet. Parents who in the past had been less actively involved in the education process will be more involved in the education of their children because of the type of information and technologies that will be available. If fact parents with Internet access at home will play the role of teachers and not rely on schools alone to educate their children.

Students will do research online and use Wikipedia, Google, Answers.com, and every available online encyclopedia or database. Schools will provide means for students to use word processors such as those available in Microsoft Office and Open Office to take notes and type research papers. Math teachers will inject the use of spreadsheets into project-based instructional endeavors, and students will be required to present their work using presentation software as is done in the real world. Information and communications technologies will eradicate most of those problems that traditional pedagogy struggled to solve.

At universities, colleges, and schools, chalkboards will be replaced by wall-mounted interactive white boards, document cameras, projection devices, and all of the equipment that makes a twenty-first-century learning environment. The textbook will be just an added resource and not the primary resource in the classroom as was the case in the past. Courses will be made available with flexibility, and access to education will be available via distance learning programs. This will eliminate the current marginalization that exists nationally and internationally between students. With high-speed Internet access, students will find available a wide array of programs, classes, and qualified professors throughout the world. This will be one solution to space and the lack of qualified teachers and will reduce the need for commuting on the part of students.

Students will register for their classes online from home, work, an Internet café, etc. Long lines at the colleges and universities to register or pay tuition will exist no more. Examinations will be given using computer-based testing (CBT), and instant feedback will be given to the student. Online instructors will give their exams on computer systems in a proctored environment. Questions on exams will be randomly generated to prevent or alleviate, if not eradicate, cheating.

Educators will have to adapt to these changes; they will have to change their instructional methods to match the changes of time. This will require training in environments that reflect the changes of time as well. The new approaches to teaching will also necessitate better pay for teachers; salaries that will allow them to be able to meet their domestic and other obligations. This will also eliminate the need for teachers to perpetrate inappropriate acts that involve accepting bribes from students for grades.

ICT will also allow educators to share lessons and materials with other educators around the world. Educational materials developed and presented in class can be stored on public servers to allow collaboration and sharing with instructors in different countries. For example, educators at William V. S. Tubman High School in Liberia will be able to share lessons with educators of Cedar Grove High School of Decatur, Georgia. Instructors at the University of Liberia will be able to share materials or ideas with professors at the University of Minnesota. An elementary school teacher at B. W. Harris School will be able to take her students on a cyber field trip, enabling them to garner the cultural diversity needed to fit into today's workplace. Educators will be able to access their grade book via the Web from anywhere in the world and make changes as needed.

Parents will be able to go onto the Internet to view their children's report cards and print them on demand because electronic grading systems will be implemented. Parents without access to computers who live in main cities or rural areas will use cell phones or smart phones to receive instant progress reports from schools.

The Liberian society will be better because of these changes that ICT will bring. Liberia will see its illiteracy rate decline; corruption, which has been a major ill of the Liberian society, will be lessened; and our human resource base will improve. Liberia will have more engineers, ICT professionals, doctors, nurses, and so on. Instead of leaving to further their studies in other countries, Liberians will engage in distance learning, while citizens of other African countries will send their workforce to be trained in Liberia. Liberians will be equipped to work in the digital economy. Farmers and "market women" or those involved in "micro-businesses" will benefit from ICT through electronic commerce; politicians will benefit by using various technologies to reach their constituents and bring development to their counties, districts, etc.

Now, let me state emphatically that I am not insinuating that ICT will be the panacea for all of the problems of Liberia's educational system. No, that is not going to happen. What I am saying, though, is that we will be able to use ICT to solve the problems of Liberia's educational system and turn those challenges that we have always been confronted with into opportunities. And using ICT to solve problems or tackle challenges does not mean that those traditional activities will simply disappear. Teachers will still have to come to school to teach despite the possibility of distance learning being implemented. Students will have to come to school and attend their classrooms and do their class work. The only difference is that they will experience a new and exciting way to learn.

To accomplish these changes, ICT will have to be incorporated into Liberia's national curricula. This means that the Ministry of Education will have to devise a plan for technology integration in schools. This plan will have to be aligned with the country's national ICT policies to ensure standardization and best practices. Also, government, the private sector, NGOs, and those of us in the Diaspora will have to make significant contributions for technological advancements to occur in Liberia. Internet service providers in Liberia will need to help schools by providing means of accessing the Internet at relatively lower costs. The following is a list of a few things that can be done in Liberia to integrate technology in schools in order to achieve a knowledge society ready to engage the global community.

- The Ministry of Education must develop a national educational technology plan, as discussed in a later chapter.

- Develop a national Web-based medium that links all schools to a repository or database of Web-based educational resources for students, teachers, and parents to access.

- Create technology education centers in local communities to allow students who attend schools that cannot afford to run their own computer labs to have the access to computer and Internet technologies.

- Set up a mobile technology education unit that will provide access to technology for schools in rural areas. This type of approach is done in Uganda, through a project known as UConnect. The project involves the use of a panel truck that carries computers that are connected to a server hosting applications that are used for e-learning to schools in rural areas of Uganda. A huge population lives in the rural sector of Liberia. This approach will help those who are being marginalized from the use of technology in education.

- Create distance-learning opportunities at local universities to alleviate the problem of students commuting from rural areas to the city in order to take classes.

- Build twenty-first-century-type schools; I discuss this in a later chapter.

These are just of few of the many measures that can to be taken to achieve our goal of providing modern education to our future generations.

Despite all of the challenges that Liberia might encounter, genuine change will come and will benefit every Liberian. The world is now experiencing the "new economy," which is based on several things: rapid technological change, increasing returns and lower per-unit costs, and increased global competition, which continues to impact prices (McConnell & Brue 2001). If Liberia is to be a part of this "new economy," there must be a shift in paradigm in every sector of the society, especially the educational sector. The road ahead promises a good number of benefits for Liberia; benefits that can only be achieved if our leaders think 70 percent for Liberia and 30 percent for themselves and not 30 percent for Liberia and 70 percent for themselves.

Chapter 5

The Need For An Educational Paradigm Shift In Liberia

In *The Road Ahead*, Bill Gates writes that "corporations are reinventing themselves around the flexible opportunities afforded by information technology [and] classrooms will have to change as well" (Gates 1995). The changes that Gates refers to are those changes that must occur in the schools to adequately equip their graduates with the requisite skills to be productive in today's workplace.

The end of the civil war in Liberia led to the genesis of a new Liberia, a Liberia rife with opportunities for development that could rival other nations currently enjoying the benefits of technological advancements and the Internet. Part of this rebuilding effort involves the building of schools and other institutions of higher learning in which future leaders of Liberia will be educated. Although the government continues to achieve strides in this area, much is still needed to be done for educational institutions to achieve the kinds of changes Bill Gates referred to. As stated in the previous chapter, Liberian schools are still being built and run using twentieth-century technology while corporations demand twenty-first-century knowledge and skills. Again, I may be redundant, but I strongly believe that if Liberia plans to compete in today's global economy, there must be a fundamental shift in paradigm, particularly in the area of education. Schools will have to be built for twenty-first-century education and be ready for twenty-first-century pedagogy, which brings me to the question that many Liberian educators had asked me during the B. W. Harris School Smart Technology Project: what is a twenty-first-century classroom?

In this chapter, I discuss the need for an educational shift in paradigm in order to be fully prepared to engage the challenges of the twenty-first century. Focus is placed on the twenty-first-century classroom.

The Twenty-first-Century Classroom

A twenty-first-century classroom involves a new type of learning; hence new teaching tools are required and used. Interactivity is one of the characteristics of this new type of learning environment. Therefore instructional materials must include new technology tools that both students and teachers can use to keep students challenged, excited, and engaged. These technology tools include computers, laptops, projection devices, document cameras, interactive whiteboards, classroom performance systems (CPSs), scanners, printers, educational software, Internet access, and so on. In a few years, schools will allow students to bring their cell phones, which will be integrated into lesson plans.

Of all the technology equipment mentioned above, the interactive whiteboard is most crucial in this learning environment. The fact that it comes with software and is touch-sensitive allows teachers to navigate with their finger or use a stylus to write content and to save the content created on the board to a computer or a repository for future reference. The saved lesson can be printed for students who may not be present in class for that exercise or can be stored at a location where it can be accessed by students and other teachers.

Another device that is used for interactivity in the twenty-first-century classroom is the classroom-response system (CPS). These devices are given to students, who use them to respond to lessons that are placed on the board by the teacher. They come with software that allows them to interoperate with the computer and screen/whiteboard and are very effective in bringing excitement to the lesson and getting students engaged. With the CPS, teachers are in a better position to assess students' activity, participation, and even attendance. Since students are assigned numbers as a way of identifying them, teachers can provide assistance and feedback to those students who struggle without their classmates knowing.

The document camera is another device that is used frequently in the twenty-first-century classroom. It comes with software that interoperates with the computer and connects to a projector to display its content on the screen. It allows teachers to scan and display traditional instructional materials, especially those materials that are not originally in electronic format, for presentation in the classroom. Materials scanned and presented can also be saved for feature reference. I always admire teachers when they use this device to teach mathematics to their students.

B. W. Harris School in Liberia adopted the document camera as part of its twenty-first-century computer lab and professional development center discussed in Chapter 8. *Not all of these devices are required in every classroom.* Some teachers are comfortable with less technology (for example an interactive whiteboard and document camera) than others are.

Apart from the technology used in the twenty-first-century classroom the teaching paradigm is another important component. As previously mentioned, the teacher in the twenty-first-century classroom facilitates learning by working with students and providing guidance on projects and other classroom activities. This allows students to fully engage with the subject matter and emerge with questions that the "facilitator" as well as the entire class can address. This also enhances critical thinking, which is a crucial twenty-first-century workplace skill.

Distinction Between The Twentieth-Century Classroom And The Twenty-first-Century Classroom

There are several distinctions between the twentieth- and the twenty-first-century classroom. The table below culled from 21stCenturyschools.com illustrates those distinctions:

Differences between Twentieth-Century and Twenty-first-Century Classrooms

20th-Century Classroom	21st-Century Classroom
The teacher is the deliverer of the lesson; what he/she says is what students assimilate.	Teacher is facilitator/coach; students work in teams and on projects.
Time-based	Outcome-based
Textbook-driven	Research-driven
Passive learning	Active learning (lots of interactivity)
Students work in isolation; no interactivity.	Students work collaboratively with classmates and others around the world. Critical thinking is enhanced.
"Discipline problems"—educators do not trust students, and vice versa. No student motivation.	No "discipline problems"—students and teachers have a mutually respectful relationship as co-learners; students are highly motivated.
The curriculum is fragmented.	The curriculum is integrated and interdisciplinary.
Students' works are assessed only by the teacher.	Assessment is done by self, peers, and other members of the audience and public.
Classroom materials are limited to texts.	Multimedia resources are used in this classroom.
Diversity is not encouraged.	The classroom encourages diversity.

Table 5.1: Source: www.21stcenturyschools.com

The Teacher's Participation In This Change

Teachers cannot be excluded when it comes to changing the Liberian classroom for the future. They must also change. They must be trained and given a basic understanding of ICT in order to adequately provide the type of education that will prepare Liberia's children for the future. Qualified teachers will always lead to better students and better citizens. The table below illustrates some of the basic skills that teachers need to have to lead a twenty-first-century classroom.

Skills and Knowledge

- Must be able to understand and use basic computer terminology
- Must be able to use computers
- Must be able to use computer operating systems
- Must be able to use word processors, spreadsheets, presentation, Web-authoring software, e-mail, browsers, databases, graphics applications, calendars, and electronic grade books
- Must be able to use computer keyboard
- Must be able to use scanners and digital cameras
- Must be able to use projection devices
- Must be able to use document cameras
- Must be able to use overhead projectors
- Must be able to design and develop multimedia products
- Must be able to use multiple data format and manipulate data

Apart from learning technology, teachers must also learn and implement other approaches in twenty-first-century pedagogy, such as project-based learning, creative problem solving, diversity, etc. In order for Liberian teachers to be effective in the twenty-first-century classroom, they will have to embrace the need to be technologically literate and must be willing to take the initiative to learn more and adapt to the changes. They will have to embrace new media that assist them in educational designs, allowing them to migrate from the "textbook and lecture style" approach.

Diversity in the twenty-first-century classroom is very crucial. Teachers must learn this and be able to provide differentiated instruction to accommodate different styles of learning. The diversity that students bring to the classroom often adds to the experiences of their classmates.

Liberian educational institutions will reach the cutting edge of educational technology in the coming years. This will only happen if our priorities are set in terms of those "basic needs" in which resources are invested. Liberia may not be able to achieve this "cutting-edge" educational technology if it continues to prioritize the acquisition of expensive motor vehicles for government officials. These motor vehicles are often driven on bad roads and soon deteriorate. Liberia can only achieve this when it invests in education that can benefit generations. And hopefully by 2015, a considerable number of Liberian educational institutions—–especially institutions of higher learning—will be able to boast of twenty-first-century classrooms.

The question of whether the country is ready to adopt this new paradigm always surfaces in conversations. When it does, I always use B. W. Harris School as proof of success. If implementation succeeded there, I see no reason why it cannot be implemented in other schools that have the potential. The problem is not that Liberia is ready; the problem is that Liberians are often too complacent with the status quo, thereby denying themselves innovation.

CHAPTER 6

EDUCATIONAL INFRASTRUCTURES IN LIBERIA MUST BE BUILT TO REFLECT THE CHANGES OF THE TWENTY-FIRST CENTURY

In May of 2009, the AllAfrica.com Web site reported that the government of Liberia had allocated $5 million USD for the construction of forty primary schools in fifteen counties. The Web site also stated that the Ministry of Education had mentioned that these new schools would be built according to "global standards of primary schools," with each school having six classrooms, a computer, and a science laboratory (allAfrica.com 2009). Upon hearing the news, I was excited and later commended the Liberian government for this initiative and for all other initiatives that sought to bring socioeconomic development and prosperity to the Liberian people (Wilkins, *Liberian Daily Observer* 2009a). The move to build schools "according to global standards," as the Ministry of Education had claimed, was and still is of great interest to me because I strongly believe that building schools to "meet global standards" entails more than the addition of a computer or science laboratory. I am of the opinion that a fully equipped twenty-first-century-ready infrastructure that facilitates twenty-first-century pedagogy is what's needed to meet global standards.

I expressed concern about this issue when I discovered that the "modern" schools being built in Liberia were mere replicas of schools that existed prior to the Liberian civil conflict (Wilkins, *Liberian Daily Observer* 2009). It appeared that the Ministry of Education was simply rebuilding twentieth-century schools and not schools that reflect the changes of our time. In a time when information and communications technology (ICT) is the driving force behind global economic prosperity, to build an edifice that lacks ICT readiness is to me retrogressive. Building schools that resemble those that existed twenty years ago when everyone has gravitated toward twenty-first-century paradigms is to me a paradox of development. This will ultimately require additional spending to upgrade those schools in the future to meet the requirements of a twenty-first century environment. This would also be imprudent and an inefficient use of tax-payers' money because their leaders were not progressive or farsighted. Liberia must be proactive and progressive in its plan to construct buildings, in particular schools, in today's ever-changing environment. This is essential to Liberia's future economic prosperity and competition in the global arena.

The amount of five million United States dollars sounds like a great deal of money to build forty schools in a third-world country that is desperately in need of infrastructure to facilitate recovery and economic development. However, to build twenty-first-century schools, five million dollars is not enough. But should desperation preclude investment in infrastructure equipped to meet global standards? To save money, would it be wiser then to build infrastructure that will not require subsequent renovations?

Most, if not all, of the schools in Liberia, be they "renovated" or "new," lack the basic infrastructural requirements that qualify them to be twenty-first-century learning environments. In light of this, newly erected schools are erroneously being referred to as "new developments." If Liberia is to erect educational environments that do not reflect global changes of today and still consider them "development," then perhaps it is time to revisit and redefine the word. But considering the amount of money allocated in May of 2009 toward building new schools, I was inclined, therefore, to believe that these schools would not be ICT-ready; and I was, without a doubt, correct.

I could not help but surmise that efforts to reduce poverty through education would be hindered by the lack of twenty-first-century-ready educational facilities that potentially renders the Liberian graduates incapable of competing in a global setting. Indeed, traditional educational facilities provided to Liberian children allow them to receive some form of education, but rudimentary at best. An education lacking modern technology in this global setting is simply limited, especially in the information age.

In a few years, the advent of broadband services will allow us to witness several changes in Liberia. Broadband companies will emerge and proliferate; ISPs and mobile operators will offer faster and converged services. Triple play bundles—voice, video, and data/Internet—will be commonplace products that Internet service providers will offer their customers, and at cheaper rates. Voice-over Internet Protocol (VoIP) will be leveraged to enhance communications because companies will take advantage of the connection to the submarine fiber-optic cable and utilize the Internet to provide cheaper means of communications. (VoIP is a networking technology through which IP voice services are delivered over broadband.) With these inevitable advancements on the horizon it is imperative that the infrastructures being constructed in Liberia today be made ICT-ready; that is, installing proper network cabling, allocating spaces for networking facilities, cable, or Internet TV facilities, etc. Whether these facilities are used now or later remains the prerogative of the tenant or custodian. But the bottom line is that the need for infrastructural adjustments would be alleviated should the need for technology integration arise in the future.

The government of Liberia has over the years invested significantly in education, which is without a doubt the best investment that can ever be made. But these investments, which are being focused on primary schools, may not provide the anticipated returns if the students are not provided education that prepares them for the twenty-first century. These students are tomorrow's leaders. If they are to have a place in Liberia's future or if Liberia will have a place in the global community, there must be a radical departure from traditional approaches to those that provide opportunities for our children in the future. Ostensibly, the plethora of schools that are being built in the country and the number of Liberians that are being given the opportunity to learn are very encouraging and an indication of the government's dedication to reducing poverty and illiteracy through education. But the problem is that our students are graduating with twentieth-century education with intention to work in a twenty-first-century work environment. And as previously stated, this kind of development is paradoxical.

Another problem this presents to Liberian students who travel abroad to further their education is the challenge of foreign schools accepting their credits from Liberia's institutions of higher learning; and I speak from experience. As a sophomore student at the University of Liberia matriculating to a college in the United States, I was under the impression that it would take at most two and a half years to complete my bachelor's degree. To my amazement and dismay, I soon discovered that several of my credits had been rejected by the independent evaluators in Miami, FL, and this influenced my decision to change my major to computer information technology (CIT) from what I was previously studying at the University of Liberia. At the time, I did not blame the University of Liberia for this setback, because I understood that it did not have the facilities or technologies that would have prevented this occurrence. But today, the availability of technologies and other resources, such as videoconferencing, the Internet, distance learning, and Web 2.0, provide many opportunities that can allow Liberian educational institutions to meet the standards of the global community.

To meet global standards, Liberian schools must have computer labs that are equipped with computers connected to the Internet, interactive whiteboards, document cameras, classroom-response systems, computers that are fully loaded with educational software, and projection devices to engage students in the learning process facilitated by teachers trained in modern instructional approaches and so on. There must also be assistive technologies available for those students who are considered "exceptional" or students with special needs.

Some of my friends and critics have raised concerns about the feasibility of implementing the changes mentioned above in absence of basic facilities and utilities in Liberia. While I do acknowledge that the absence of these basic necessities—especially in remote areas—might delay implementation of these changes, my argument is that Liberia, at the very least, must leverage what is currently available or find innovative ways to adapt to the changes of the twenty-first century. I know full national development will take some time to be achieved, but the work can begin in Monrovia first and gradually move to remote areas as we make improvements. If we do this, we should not have to play catch-up in the future.

Finally, investments in education must also include ICT in order to support the government's Poverty Reduction Strategy. Development must involve those initiatives that will prepare Liberia for the global setting not simply for the sake of saying that we are making progress. Times have changed; people have changed; even mission statements, according to Bill Gates, have changed. Hence Liberia must change to prevent being left behind as it has been in past years.

The world is willing to help Liberia, and therefore Liberians should be willing to help themselves. Liberians need to take advantage of the assistance provided by the international community and use it for the modernization of the country. It is time we divorce ourselves from those initiatives that make development in Liberia seem "paradoxical." When Liberia finally engages and implements those approaches that will bring modernity, I believe that will be the genesis of Liberia's exodus from poverty and illiteracy to a prosperous and knowledge-based society.

CHAPTER 7

DISTANCE LEARNING IN INSTITUTIONS OF HIGHER LEARNING

Whenever I travel to Liberia, I often visit the University of Liberia's main campus. However, my visit to the campus on January 6, 2010, was very different. I believe it was during this visit that it became apparent to me that the university had not changed; it had not improved. It was reminiscent of my time spent there as a student during the late '80s and early '90s. Overcrowded classes, classes where students sat entirely too close to the instructor while others stood outside by open windows to take notes, could still be seen. I recall, while walking toward the second floor of Tubman Hall, saying to myself, "We are in the twenty-first century; this should not be happening." In contrast, the university's Fendell campus, located outside of Monrovia, is currently being renovated by the Chinese government with the intention of conforming to global standards. However, for some reason, the conditions and standards of the main campus remained dilapidated and antiquated.

This chapter discusses distance learning as an alternative and solution to the problems faced by both students and instructors at universities in Liberia. The recommendations that I make here may not all work in Liberia since ICT penetration is still in its infancy. But there are prospects that lead me to believe that this form of learning will soon infiltrate the Liberian educational system with great success.

As an ICT professional I am charged with the responsibility of using innovative information technologies to facilitate processes and make life easier for stakeholders. With that in mind, I thought the situation at the University of Liberia's main campus and other Liberian institutions of learning required an IT-related solution. I concluded that the implementation of a distance-learning program might ease the struggles that students and professors experienced daily in their academic pursuits. I was cognizant of the potential challenges this change would bring because distance learning would be a shift in paradigm and a new culture. I was prepared to hear arguments against distance learning by instructors who might be resistant to change and comfortable with the status quo. But progress cannot be deferred or delayed for a select few. Instructors will need to embrace change in pedagogy and the understanding that we have a responsibility to prepare the country's future. If change is to occur, they too must adjust.

Distance learning is an educational setting in which there is a geographical distance or separation between the student and the instructor. It allows course availability, especially in situations where infrastructure is limited or simply not available. Distance learning can alleviate problems that students and instructors currently face at the University of Liberia and other institutions, such as space, commuting cost, qualified labor, etc. It will increase access to education and will improve learning because of the opportunity to interact and collaborate with a diverse group of instructors situated in different countries.

Implementation of distance learning in Liberia can be done at a reasonable cost because of advances and progress made in technology and telecommunications, mobile technologies in particular. Distance learning without broadband connectivity is feasible but may include some challenges. Albeit very cost prohibitive, satellite communications technologies can be used to deliver distance learning. A low-cost pre-broadband-installation distance-learning program can be implemented using free and open source software like Moodle. Moodle is course-management software that can be used in place of costly proprietary solutions such as Blackboard or WebCT. Obviously training and technical support regarding the use of distance learning software is required for instructors and students. However, institutions opting to use proprietary software should negotiate vendor-sponsored training.

Types Of Distance-Learning Technologies

There are many types of distance learning programs and technologies. Online learning, as described above, is only one form of distance learning. Below, I discuss additional ways distance learning may be implemented according to the University Wisconsin (Wisconsin n.d.).

Audio conference—This type of distance learning includes the use of telephones to connect the instructors and students. Would this form of distance learning be a good fit for Liberia at this time or when broadband reaches the country? The ubiquity of mobile phones provides a medium for this option. However, the cost of purchasing phone cards is a potential deterrent. But with the progress that continues to be made in the field of telecommunications, there is great possibility that the cost of phone calls will decline and audio conference distance learning will be considered.

Online distance learning—This type of online distance learning involves the delivery of courses over the Internet, and all activities are Web-based. It involves the use of course-management systems such D2L, WebCT, Blackboard, the open source Moodle, etc. Course-management software is often used in online learning to organize content, activities, communication, and assessments. Some course-management software used in online distance learning comes packaged with the following: chat room environments, discussion boards, grade books, storage facilities, e-mail access, and instructional resources for both students and instructors. Online distance learning allows students in different parts of the world to simultaneously participate in the same virtual classroom. It is by far the most widely used form of distance learning.

The University of Liberia and other colleges and universities in Liberia will need to adopt online distance learning to expand their academic programs. I would suggest using the open source Moodle course-management system initially since it is cost effective. Another option would be to use Web 2.0 via social networks for distance learning. Colleges should begin this endeavor in their professional development or graduate programs prior to transferring it to other programs. Web-based solutions will facilitate online distance-learning penetration into the Liberian education system. With mid-range servers, enough storage, faster processors, and adequate bandwidth, these institutions of learning can implement online learning after providing instructors training in online instruction.

Multimedia—This type of distance learning would be a good fit for Liberian universities and colleges because the course materials are stored on CDs, DVDs, flash drives, videocassettes, audiocassettes, or any onc of the available stored media. The courses presented via multimedia combine texts, graphics, audio, video, etc. Some multimedia courses require Internet access, but most of them do not. Most of all, multimedia courses are specifically designed for flexibility, modularity, and to run at the student's pace. This type of distance learning will be best used for professional development in government, NGOs, businesses, institutions of higher learning, etc.

Print—The print distance-learning option is another distance-learning program that can be implemented in Liberian schools. Several steps will have to be taken to ensure the integrity of the program, especially when it comes to assessment. Print distance learning is done through the postal system. Course packets and other educational materials are mailed to students using the postal system. The challenge associated with this type of distance learning in Liberia is that currently the mailing system is neither efficient nor reliable to ensure on-time delivery of content to students. Students who choose this type of distance learning submit their work via the post office or any postal service, by fax, and sometimes electronically via e-mail. Assignments and exams are self-paced and submitted within the specified timeframe. I am not certain when Liberians will adapt to this form of distance learning due to the mailing issues previously mentioned. I am confident though, that when those issues are addressed, institutions will gravitate toward this option.

Webcast—This option will only be feasible once broadband connectivity is available in Liberia. Webcast is the process of capturing and recording multimedia materials and other types of digital data and then synchronizing it as a single streamed media presentation (Wisconsin n.d.). Webcast courses can be viewed either during real time or later by accessing the Webcast's link. Interaction between instructors and students is done in a variety of ways, including chat rooms, discussion and message boards, e-mail, scheduled audio conferences, etc.

Telecourse/datacast—The proliferation of television stations in Liberia and the frequency with which they broadcast makes this form of distance learning likely to occur much sooner than other forms of distance learning. Telecourses are highly produced videotaped courses that are broadcasted at scheduled times by television stations. Study materials are given to students along with assignments and directions. Datacasting, on the other hand, involves the transmission of multimedia materials via airwaves along with the digital television signal. Content created through datacasts are made available for download to a computer or to be viewed on a television (Wisconsin n.d.).

Videoconference—Videoconferencing is becoming popular these days because of the ubiquity and flexibility of Webcams, which has made what once was considered an expensive setup an affordable one. Although Webcams do not make for the best videoconferencing, they provide a more convenient and portable means of taking courses. Videoconferencing allows a synchronous two-way communication between the student and the instructor, allowing real-time discussions (Wisconsin n.d.).

Videoconferencing will be used by colleges to extend their academic program in Liberia, especially if Webcams and other low-cost technologies can be integrated for better content delivery. While I expect colleges to use this option aggressively, I strongly believe that the government of Liberia will also benefit if it were to utilize videoconferencing for communications, especially with superintendents and other leaders in rural areas. This will reduce travel and its associated costs.

Web conference—Whenever I am asked about Web conferencing, I often use videoconferencing as an analogy, although they differ in many ways. Web conferencing is done over the Internet because it utilizes both a Web browser and an audio conference to enhance interactivity. It brings a multitude of individuals together from different locations, creating an interactive environment for the sharing of information. It is used significantly in academia, business, and government environments in the United States. Several things can happen during a Web conference: students can share documents and send and receive text, graphics, etc. Web conferencing may not be an ideal solution for Liberian academia at this time since much of Liberia's technology initiatives are still in the infancy stages. However, with broadband connectivity, schools that are innovative will gravitate toward Web conferencing as a means of cutting the cost of their distance-learning programs.

Advantages And Disadvantages Of Distance Learning

While distance learning has become a new way of expanding education, it, like all other initiatives, has its advantages and disadvantages. Below is a discussion of both.

Advantages

Flexibility never seen before—One of the advantages of distance-learning courses is that they give students flexibility—the ability to do school work from anywhere in the world at any time—with the only prerequisite being a computer with Internet connection. Now, this applies only to those distance-learning programs that require such equipment. Some distance-learning programs, as mentioned above, may not require Internet access. A typical example of this is the print distance-learning option. The fact that students have this flexibility allows them to have both a professional and a personal life as compared to the traditional format, which does not permit both or makes having both difficult.

At the University of Liberia, which lately has been overcrowded, a distance-learning program would be ideal. It would give students another option to take the classes needed to graduate. Obviously, not all students will opt for distance learning, but for those students who would, having it available will prove, ultimately, beneficial to the school, the student, and the country.

Also, the proliferation of Internet cafés and the advent of broadband in Liberia will influence many university and college students to gravitate toward online distance learning. This will be extended throughout the country when the cost of Internet connectivity drops and universities and colleges move toward integrating mobility (offering courses via mobile devices) to allow broader participation of students. Visibly, mobile technology in Liberia has had a larger penetration than information technology; hence the potential for it to succeed in Liberian academia is greater.

Students do not have to commute—As a student of the University of Liberia, one of the main troubles I faced was with transportation. Since I could not afford the luxury of owning a vehicle, I relied on public transportation to commute to and from school. Public transportation has always been a major problem in Liberia. Universities could alleviate this problem by implementing distance-learning programs. College students could then take advantage of this flexible learning format. The days of standing in the rain, waiting in lines, and struggling for seats in classrooms would soon be gone.

Availability of courses—Overcrowded classes, which are a direct result of limited number of instructors or the unavailability of courses, could also be addressed via distance learning. It would allow students in Lower Grand Gedeh County who are interested in taking a particular course that is currently not being offered or is too full to take it either online or through an optional distance-learning program.

While a student at the University of Liberia, I did experience the frustration of not being able to take a class because of the lack of instructors. At that time, however, one's only recourse was to wait until the next semester to register for the course. A consequence of this situation was delayed graduations. Distance learning in Liberian institutions would allow professional and educated Liberians in the Diaspora who are willing to impart knowledge to their fellow Liberians to work as distance-learning instructors.

DISADVANTAGES

No physical interaction—One main disadvantage of distance learning is the lack of physical interaction with the instructor and your classmate(s) as is found in the traditional classroom. To minimize this isolation, today's distance-learning programs have attached new communications technologies such as chat, discussion board, e-mail, conferencing, bulletin boards, or a one-weekend-on-campus requirement.

In the case of Liberia, which is a small country, not seeing your classmates may not be an issue. There is the possibility that students will see each other more frequently than they would in a larger city or country. Students can meet at libraries, their classmates' homes, and other venues. Now, this may not be the case if the institution establishes an international or regional distance-learning program in which students and instructors are spread across the global spectrum.

Cost—Another disadvantage that impacts distance learning is the cost of implementation, especially those forms of distance learning that require high-tech equipment such as videoconferencing equipment, computers, and so on. Students who choose the online distance-learning option are required to have a computer with Internet access, which can be quite expensive. But those students who opt for this method of learning would probably be in a position to do so.

Technology—Most of the distance-learning programs listed above require some sort of technology implementation—a computer, Internet access, videoconferencing device, etc. The equipment is costly and may be a factor that deters many institutions from adopting those kinds of distance-learning programs in Liberia. Students, on the other hand, may not be able to afford to purchase new computers or other resources needed in a distance-learning program. Also, knowledge of computers and their uses is required. This means that colleges offering distance learning that require extensive use of computers and the Internet will have to provide classes that will equip students with basic computers skills.

Self-motivation and advance planning—Distance learning is not for everyone. It is only for those who are self-motivated and good planners.

Theoretically, this should not be a problem for Liberian students, especially students attending institutions of higher learning, because typically they are quite mature and determined. However, the kind of discipline expected for success in this type of program is different, and a student who would be stellar in a traditional setting may not be in an online setting. Also required in distance-learning programs are excellent organizational skills for both the student and the instructor.

The advantages and disadvantages discussed above are not exhaustive. But I would expect that any implementation of distance learning would first benefit from thorough research and understanding of all the pros and cons.

Distance Learning's Impact On Instructors

Distance learning will be a good tool for professional development in Liberia, especially for those instructors who continuously seek opportunities to learn. As it gains popularity and reaches maturation, distance learning will be used for human-resource development by employers. Liberian teachers will benefit from this as many schools, both public and private, will engage in such training. The Ministry of Education, which sets standards for instructional practices in Liberia, will move toward distance-learning programs to provide training that may not be available at Liberian colleges or universities in traditional format.

Teachers and students residing in Maryland County will be able to take courses at the University of Liberia or Cuttington University College without having to leave their place of work or residence. Overall, quality education will be provided, transportation costs will be reduced if not eliminated, and Liberia will ultimately be able to achieve the status of a knowledgeable and literate society with the capability of competing in the global community. Distance learning will present some challenges initially, but ultimately it will be adopted and embraced by the Liberian community.

Chapter 8

Pioneering Twenty-first-Century Educational Technologies In The Liberian Classroom: The Case Of The B. W. Harris School Smart Technology Project

Before the B. W. Harris School started on the path toward the information age, I can recall touring the school in 2005. I noticed a storage room full of antiquated computers that were probably donated but—for some unknown reason—appeared to never have been used. Naturally, I inquired from some of the teachers the reason for the storage and non-use of the computers. I was informed that they were used initially in a typing class and then abandoned because they were not functional. And sadly, the teachers knew of no one who could repair them. This year marked my first trip back to Liberia after a protracted period of absence and dreams of making a dramatic return.

To garner more information, I polled a cross-section of students at the school to determine their interest, if any, in a new way of learning that would prepare them for twenty-first-century careers. As I listened carefully to each student, a feeling of nostalgia enveloped me and I began to reminisce over my days as a student there. Then I had a vision of a new and improved B. W. Harris School. By the end of that fateful visit, my thoughts ran to the whole of Liberia, one with a myriad of opportunities and a new "lease on life"; its potential to be a developed and economically viable country in West Africa. It was then that I concluded that Liberian schools were in desperate need of a shift in pedagogical paradigm; that our students were still lagging behind the rest of the world because of obsolete educational resources that did not reflect the changes of the twenty-first century.

With the exception of the cell phone, which every student had, there was a lot that Liberian students did not know about the global community and the impact technological changes had on it. The mere fact that students were being given an education that had the potential to be dismissed by evaluation firms or colleges in Western countries bothered me greatly. This became a defining moment, a moment when I decided to divorce myself from the fear of a departure from the stakes (family, home, job, friends, democracy, security, etc.) that I had left behind in the United States to return to Liberia and make a difference; a difference that would impact generations to come.

And so, with knowledge of the Liberian educational situation and a renewed enthusiasm, dedication, and empathy, I returned to the United States fully convinced that I would return with something that will bring change; change that will help alter the course that education was taking in Liberia. This initiative was not just to pioneer a modernization of the educational system but an attempt to help in the process of preparing Liberian students to work in the twenty-first century.

In the United States, B. W. Harris School has a strong alumni association that provides needed support to the school in Liberia. So upon my return, I made a full report of my findings to the then national chairman of the association, Miss Wilhelmina Wilson. In my report, I mentioned the school's need for modern technology and how the current setup would not be able to support the integration of the twenty-first-century paradigms I'd envisioned. Miss Wilson acknowledged my findings, but because it was her last term, understandably, she was unable to act. She did, however, mention that she would work with the new administration of the association to ensure that the school integrated modern technology. Interestingly, I would be a member of the newly elected administration and one of those to further the "change" that I had envisioned.

In July of 2007, I was elected second national vice chairman of the B. W. Harris Alumni Association, USA. Mr. Handel Diggs and Mr. Moses Okai had been elected national chairman and first vice national chairman respectively, although Mr. Okai had to be replaced by Miss Eugenia Jelani since he (Mr. Okai) had to relocate. Also elected were Mrs. Yvonne Hansford, Mr. Stephen Taylor, Mrs. Famatta Zeon, and Miss Eudora Gardiner. Each of these individuals would play a major role in what were to be several pioneering initiatives that would alter the way students learned at B. W. Harris School forever.

The night before our elections into office, I had mentioned to Mr. Diggs that I had a plan for the school. I briefly informed him about the computer lab, which he endorsed unofficially. He, however, encouraged me to put it in writing and forward it to him so that he could present it to the board of advisors. After our election into office, I presented a proposal to Chairman Diggs, who included it on the agenda for the next board and general assembly meeting. This was in July of 2008.

At our first board meeting, I was asked to present the proposal. This proved to be a problem because I had lost everything I was to present when my laptop's hard drive failed. Without the presentation, I tried to present the proposal extemporaneously. I think I caused more confusion than clarity that day, because only a few questions were posed during my presentation. The following day would be the general assembly, at which time there was a blitzkrieg of questions that I struggled to answer. Fortunately, Counselor Mohamedu F. Jones, a prominent member of the association and the Liberian community in the United States, rose up and proclaimed that "this is one of the best projects the association has ever embarked on" and that it should be approved; and it was!

In December of 2008, I was sent by the B. W. Harris School Alumni Association in the United States to implement the Smart Technology Project. I arrived in Liberia to a warm welcome by the school and the local alumni association. The project was embraced prior to my arrival, and all of the teachers looked forward to what was to come. This marked the beginning of the aforementioned shift in paradigm.

The Smart Technology Project

The overall goal of the Smart Technology project was to bridge the technological and educational chasm that exists between the students of developed countries and Liberia. We wanted to change the instructional paradigm to use those approaches that would prepare the students for the twenty-first-century workplace or the digital economy. We wanted to transform our alma mater from a traditional educational environment to one that parallels a twenty-first-century learning environment, which is characterized by the use of technology for collaborative learning. In doing so, we dispatched a consignment of computers, network servers, LCD projectors, camcorders, audiovisual equipment, and other technology-related equipment to be deployed at the school on Broad Street, Snapper Hill. The project, code-named "Smart Tech" by Chairman Diggs, would introduce a new pedagogical paradigm that would enhance creativity, innovation, technology literacy, national literacy, and global literacy.

The Smart Tech project was designed to be implemented in three phases. Phase I is described below; phase II was the implementation of a professional development center; and phase III would include an electronic grade book system, digitized instructional materials, and so on. Below is a list of the tasks that were completed in phase I of the project:

- Installation of a twenty-first-century-like computer lab with twenty-five computers, an LCD projector for presentation, Internet access, scanning, and printing capabilities. Also included in the computer lab was interactive educational software to enhance student learning.

- Converted the library to a media center, making it a site that provides *limitless* access to information. Installed computers with Internet access in the library for students to use for research purposes; provided audio/video technologies with a library of educational videos on VHS, CDs, and DVD for visual learning as well as video-production capabilities.

- Developed a Web presence for the school as a one-stop shop for information to enhance cultural and scholarly exchange with other students around the world.

- Developed a mini computer lab using desktop virtualization to increase student-to-computer ratio.

- Formed a team of students to advance technology at the school. This team was named the SWAT Team (Students Working to Advance Technology), and it consisted of ten students. The SWAT team was trained to install and maintain computer systems, network administration, network security, Web design, and basic computer programming.
- Trained teachers on how to integrate technology into the school's curriculum, collaborative learning, word processing, spreadsheet, and presentation using Microsoft Office 2003 and differentiated instruction.

During phase I of the Smart Tech project, we held a nine-day technology seminar for teachers. The seminar was intended to train teachers on how to integrate technology into their classrooms. It gave the teachers the basic understanding needed to familiarize themselves with the twenty-first-century technologies and instruction. I strongly believe that a well-trained faculty has the potential to ensure that present and future generation of Liberians will be technologically savvy and prepared to take on the challenges of the twenty-first century. Topics presented during the teacher's training sessions included introduction to computers, integrating technology into the curriculum, and the twenty-first-century classroom. Teachers were trained on how to use the Internet for teaching and learning, Microsoft Office Suite with emphasis on Microsoft Excel for grade book purposes, and Microsoft PowerPoint to present lessons in the classroom. The training also covered collaborative learning and teaching. Teachers learned how to set up technology equipment such as computers, projectors, document cameras, etc., in their classrooms for instructional purposes.

Also involved in the training was the SWAT, ten students from the ninth through the eleventh grade, who received the following technical training: introduction to computers, building and maintaining computers, network administration, network security, Web design, and basic computer programming. Upon completion of the training, the SWAT Team would serve as technical support aides for the school. Their tasks then and now are to provide technological assistance to their fellow students and work toward advancing technology at the school.

This was an effort to give students the opportunity to learn skills that can be applied in the workplace upon graduation.

The SWAT program parallels the type of employment skills elective classes or extracurricular programs in the United States impart to their students; these skills will be a great asset to Liberia if the schools begin to incorporate a similar paradigm into their curricula.

Smart Tech Phase II 2009–2010

Phase II of the Smart Tech Project was packaged with yet another developmental initiative. These initiatives included project "Spartan Pride," which involved a $22,000 renovation of the school; a $17,000 teachers' salary supplement; the construction of a high-tech professional development center; twenty academic scholarships for students who met the required criteria; and a teacher-training boot camp that began December 21, 2009, and ended December 23, 2009.

Project "Spartan Pride" was a major initiative undertaken by the association to renovate the school. It was the result of an earlier assessment report that highlighted the school's needs that was submitted by Mr. Anthony Deline II, the then interim administrator at the school.

The most controversial aspect of the association's 2009 projects was the new twenty-first-century professional-development center and the teacher-training boot camp. The professional-development center was built to provide teachers and staff with professional-development seminars and programs. The center was equipped with modern technology, including a computer with Internet access for presentations, a mounted LCD projector, an interactive whiteboard, a document camera, a video player system (DVD/VHS combo player), an overhead projector, a classroom audio system, and so on.

Upon completion of the center, we held a three-day professional development "boot camp" to give teachers a preview of the slated pedagogical changes. Ten schools in and around Monrovia along with several professionals organizations in the area of education from NGOs, higher institutions of learning, and the Ministry of Education were invited to do a series of lectures and presentations.

The boot camp for teachers was a success. It began with a formal opening that brought in an official from the Ministry of Education, the press, and the other invited participants. During the ceremony, we gave a demonstration using the interactive whiteboard and the AVerVision document camera. That demonstration generated a lot of questions and concerns about whether the teachers would be able to use their new equipment. Initially, I too was concerned; I was not certain if only three days would provide them with the requisite skills and a good level of comfort. But I was wrong; the teachers amazed me with the way they quickly embraced and mastered the use of the equipment. In fact, as they became more comfortable they began to request more. All of a sudden it was about supply and demand!

During the boot camp we had teachers work in teams, listen to lectures, role-play, and perform hands-on activities. The training gave the participants knowledge of twenty-first-century instructional approaches, which gave them the competence required to impart quality education to the students using a different approach. This new approach will create a new level of enthusiasm in students to learn and prepare themselves for standardized tests. It will also instill in them those skills that will enable them to confidently navigate the twenty-first-century work environment and, most of all, prepare them to compete in the global economy.

Highlighted during the Smart Tech phase II and the teachers' boot camp were the school's newest technology equipment—the AVerVision document camera, which teachers were trained to use to digitize manual instructional works. Additionally, the interactive whiteboard—created using a regular whiteboard and Mimio interactive device—was introduced to the teachers. This device allows teachers to use a stylus (a pen-like instrument) instead of chalk. The overall goal was to replace the traditional blackboard and chalk with the interactive whiteboard and stylus. Both the interactive whiteboard and the AVerVision document camera were intended to allow teachers to create instructional materials in digital format and store them on the school's server for future use.

Future uses include providing a repository for students to retrieve instructional resources when they miss school or posting instructional resources on the Internet to enhance international collaboration with other educators. With those instructional technologies in the classroom, teachers would then be able act as "facilitator" as is done in the quintessential twenty-first-century classroom, allowing students to work in teams on projects to enhance critical thinking.

The Smart Tech project also brought Internet access to the school as well as the implementation of desktop virtualization, a relatively new technology that allows multiple computer monitors to display the contents of a single computer.

Being the leader of the Smart Tech project, my vision for B. W. Harris included converting classrooms to twenty-first-century-like classrooms; providing regular professional development for teachers on twenty-first-century instruction; installing an electronic grade book system to replace the current system, which is done manually; installing a school-to-home communications system to update parents on students' progress; allowing teachers to digitize their lecture notes and store them on servers for future use; allowing students to use the Internet to participate in global and virtual classrooms, etc. Initially, for some, it sounded impossible and too bold a project to implement at a school situated in a country recovering from a prolonged civil war. But the proof is in the pudding—it can be done!

What B. W. Harris Alumni Association, USA, has done for its alma mater is to pioneer a new educational paradigm. By installing twenty-first-century classroom technology, the organization has brought a change that will ultimately be emulated or adapted by other schools in the country. The new approach will enhance critical thinking, which has not been a major part of learning in Liberian schools. The hope and dream is that this new approach in education will spill over into the whole Liberian school system—public and private—but this will require work and support from local Liberians and those in the Diaspora.

CHAPTER 9

HOW LIBERIAN ACADEMIA CAN BENEFIT FROM THE MASSACHUSETTS INSTITUTE OF TECHNOLOGY (MIT) OPENCOURSEWARE INITIATIVE

Twenty-first-century technological innovations involve a great deal of collaboration. We have seen several examples of this collaboration at work: YouTube, Wikipedia, Linux, etc. Collaborative work is not a new concept by any stretch of the imagination; in fact, it has been an integral part of academia for many years now. It has increased access to educational resources for many individuals throughout the world despite their geographical location. An underutilized and often overlooked source of collaborative work that offers great benefits and incentives for both students and instructors alike is the MIT OpenCourseWare initiative. MIT OpenCourseWare could positively impact the educational sector in developing countries that lack resources that needed to garner quality education. In Liberia, a quintessential developing country, it will provide students, instructors, and professionals with access to study materials used at one of the world's greatest technical universities, the Massachusetts Institute of Technology.

It would be redundant to reiterate the impact that technology has had on mankind; however, sharing with you the impact technology has had on making my dreams become reality is not. At least not to me!

As a young man in Liberia, my dream was to pursue an education at the Massachusetts Institute of Technology. And having always been an honor student throughout high school, my dream was certainly possible. But then the Liberian civil war came and my parents lost everything!

Years later, as graduate students, my classmates and I found ourselves using syllabi, lecture notes, and projects from the Massachusetts Institute of Technology (MIT) even though we were not students of the prestigious institution. Not only were we allowed to share educational materials and engage in intellectual discussions with students at the institution, but access to these resources was free of charge. This endeavor was an attempt to familiarize ourselves with the quality of an MIT education. I also used this opportunity to investigate the possibility of integrating this resource into the Liberian education system.

So what do I hope to accomplish in this chapter? It is my hope that leaders in Liberia's institutions of higher learning will consider the OpenCourseWare initiative as a supplement to their current pedagogical approaches and engage in activities similar to those of MIT. By engaging in similar activities, I mean the creation of a Liberian OpenCourseWare initiative where the works of students and instructors are placed in a central location—on a server—for other students to use as a resource. This is the same approach that we began at B. W. Harris School even though B. W. Harris's initiative was not an open one but rather an internal attempt to facilitate student learning at the school.

MIT OpenCourseWare (OCW) is a web-based publication of virtually all MIT course content in a single location—a server—on the Internet (OpenCourseware n.d.). This is a result of the faculty at MIT's passionate belief in its mission, which is based on the philosophy that "open dissemination of knowledge and information can open new doors to the powerful benefits of education for humanity around the world (OpenCourseware n.d.). It offers free and open access to educational materials from its undergraduate and graduate courses, spanning thirty-three of MIT's academic disciplines and all five of its schools, including the School of Architecture and Urban Planning, the School of Engineering, the School of Science, the School of Humanities, Arts, and Social Sciences, and the Sloan School of Management. It is "open and available to the world and it is a permanent MIT activity. MIT OpenCourseWare offers a free publication of MIT course materials that reflects almost all of the undergraduate and graduate subjects taught at the institution" (MIT 2005). Yet there are a few things about MIT OpenCourseWare that potential users need to know. First, OpenCourseWare is not an MIT education! This means that using these resources does not make you a student of MIT. Second, MIT OpenCourseWare does not grant degrees or certificates to those who use its resources (OpenCourseware n.d.). While you are given an opportunity to utilize the resources that are available at MIT, you are not considered a student of the institution, and therefore you are not entitled to a certificate, diploma, or degree. Third, using MIT OpenCourseWare does not give you access to the faculty of MIT, and they are therefore not required to respond to your questions or concerns. Remember, you are only benefiting from the academic resources used at MIT, not necessarily its personnel. Last, MIT OpenCourseWare and its materials provided may not reflect all the contents of the courses taught at MIT. There may be materials that professors or the institution choose not to make available. Hence, I would advise a thorough analysis of materials accessed from MIT OpenCourseWare prior to presenting it in your classroom if you are an instructor. This way, you as the instructor may search elsewhere if you need more materials to buttress your presentation in class.

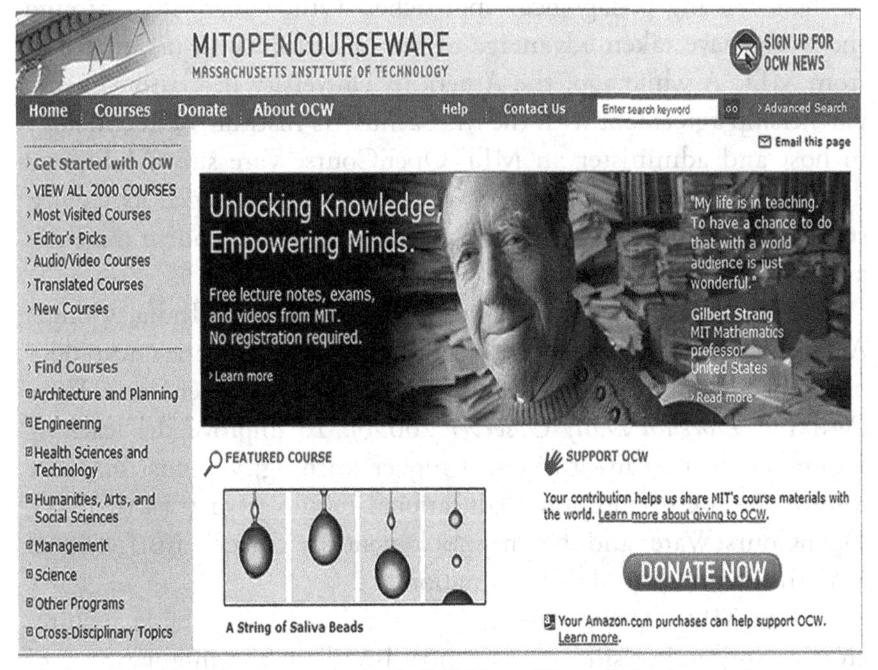

Figure 9.1. A snapshot of MIT OpenCourseWare's website

As I mentioned earlier, while a student, I personally benefitted from MIT OpenCourseWare. I took two classes for an entire semester without purchasing the required textbooks because the textbooks were too expensive and my professors primarily referenced current issues and articles related to the industry. According to the professors, cases in the textbook were "antiquated" in the world of information technology. Instead they used real-world experiences and case studies to encourage a new learning approach to problem solving known as peer collaboration. Peer collaboration involved a diverse group of students working together on virtual teams on a particular project that aimed at bringing new innovations and improvements to existing processes. The end result was usually a product or idea that could be submitted for publication in *Association for Computing Machinery (ACM)* journal or to large companies for further improvement through research and development for consumers.

In Liberian academia, MIT OpenCourseWare would be a great resource because of the demands by students for educational resources and textbooks. These institutions could begin to explore resources on the MIT OpenCourseWare Web site and utilize them as supplements to their existing pedagogical approaches. Other universities around the world have taken advantage of the opportunity to use materials from MIT. A while ago, the American University in Cairo signed a partnership agreement with the Massachusetts Institute of Technology to host and administer an MIT OpenCourseWare site (Mahmoud 2009). The American University in Cairo also encouraged its faculty, students, and researchers to make use of the online course materials available on their servers.

Other countries such as Ghana, Cameron, Iran, India, Congo, Algeria, Ethiopia, Kenya, Mali, Nigeria, Senegal, and Liberia's next- door neighbor—Cote d' Ivoire—are also partners of MIT OpenCourseWare (Wilkins, *Liberian Daily Observer* 2009b). To improve the learning system at Liberian institutions of higher learning, we must gravitate toward the use of additional educational resources such as the MIT OpenCourseWare and begin supplementing current instructional initiatives with up-to-date information.

Using MIT OpenCourseWare at a university in Liberia requires creative approaches since the country is still in the infant stages of technology infusion in education. My first suggestion would be a reiteration of what I suggested above; that is, to emulate those countries that have partnered with MIT OpenCourseWare. In partnering with MIT OpenCourseWare, universities and colleges in Liberia will have the opportunity to run their own servers and be responsible for some of the tasks involved in publications. They will also be responsible for the delivery of materials to students and researchers. This would be a better option because it keeps the college in close contact with OpenCourseWare material. Alternatively, universities could invest in large printers or photocopy machines that could be used to print materials from MIT OpenCourseWare to distribute to students free of charge. The next alternative could be to provide computers with Internet access to students to allow them access to OpenCourseWare. But the overall goal will be to use OpenCourseWare materials as a supplement to existing resources. This will allow students to gain access to updated educational materials.

ICT and other training institutions that may be in need of up-to-date instructional materials would now have access to MIT OpenCourseWare materials as well. This will allow them to provide better curricula. Now, while OpenCourseWare does not provide the practical training that is required for some ICT training programs, some of the courses offered can be used to broaden students' knowledge of information and communications technology. The following are a few courses provided by MIT OpenCourseWare that I would recommend to "computer schools" in Liberia: Introduction to Computer Science and Programming, Introduction to Software Engineering using Java, Principles of Digital Communications, programming languages, Principles of Computer Systems, Network and Computer Security, Introduction to Copyright Law, Introduction to Technology and Policy, Communications and Information Policy, Database Systems, Network Systems, etc. These are just a few of the courses offered by OpenCourseWare that I believe can be offered at computer schools in Liberia without the need for hands-on materials. I would place more emphasis on computer programming and software engineering, because as we move ahead, programming and software design skills will be greatly needed. Engaging in software development at this time will be the best way to prepare for the twenty-first-century information highway, which, according to Bill Gates, will rely on software overwhelmingly (Bill Gates 1995).

On the other hand, business schools can also take advantage of this opportunity by using materials from MIT/Sloan School of Management. A management program in Liberia can use as a resource the following courses from MIT OpenCourseWare: Operations Management, Economic Analysis for Business Decisions, People and Organizations, Introduction to Financial and Managerial Accounting, Managerial Psychology, Integrating eSystems and Global Information Systems, etc. All of these materials are available through MIT OpenCourseWare free of charge.

Although the use of resources from MIT OpenCourseWare may not necessarily level the playing field between a student of computer science in the United States and one at the University of Liberia, it will give the student at the University of Liberia a slight competitive edge when placed in the global setting. There is much that needs to be done to improve the quality of education in Liberia and bring it up to the standards of other schools in the global community. But to get to where we want to be, we have to consider even the most infinitesimal contribution to the improvement of education. As I always say, Liberians will have to be creative and find innovative ways to solve difficult problems. The use of MIT OpenCourseWare is a creative solution to improving higher education in Liberia.

Incidentally, students attending institutions of higher learning, especially students at the University of Liberia, must not rely solely on their professors for knowledge (in fact, no student should do so). They must engage in personal endeavors that can help expand their horizons in their chosen academic discipline. The Liberian notion of "I don't need to learn too much because when I get the job, my special assistant will take care of rest" will never work in the new and global economy. Students must understand that the education they garner from institutions of higher learning is what they will carry into the global setting. Failure to do good and honest work while in school will simply make you, as my late father often said, an "uneducated person with an education." That saying refers to individuals who graduated from college but did not have the capacity to engage in any intellectual discourse that related to what they claimed to have studied in school.

If for some reason your university is not able to provide up-to-date materials and resources for your classes, you should take the initiative and explore other options. Remember that we have the Internet, and while I understand that Liberia may still be transitioning in terms of its infrastructure, there are Internet cafés and other means of accessing the Internet for scholarly materials. I fully understand that times are hard and many university students do not have jobs to support themselves and their families. And this is why I advocate the use of MIT OpenCourseWare as an alternative to expensive or unavailable books needed for educational initiatives in higher institutions of learning in Liberia.

Chapter 10

Why Liberian Students Should Pursue Careers in ICT—The Path To Success In A Digital Economy

As an ICT professional, I spend most of my free time consulting and guiding individuals on issues relating to technology. Oftentimes, I consult individuals who have issues with or who are in the process of planning their systems. There are times when I get calls from family and friends about computer hardware problems—I really do look forward to these calls. They are always either amazing or very interesting, and I try to help in any way that I can.

One of my many "extracurricular" activities is advising and providing guidance to Liberian youth who aspire to be ICT professionals, entrepreneurs, etc. Since being a "mentor" to a few young Liberians, I have discovered that there are many young high school seniors and college freshmen who desire to be ICT professionals but have been discouraged because of the country's slow progression in the area of ICT. In this chapter, I take a modest approach to educate the public about careers and opportunities in the fields of computing and ICT. I purposely cover only a few careers, because an attempt to include all would be unrealistic at best.

After over ten years of chatting in many Internet chat rooms, one conversation still resonates with me. It was a chat that I had with a sixteen-year-old Liberian student who happened to be my nephew. He had been contemplating his future career and thought about ICT. He had written me several e-mails prior to our Internet chat, and in one he wrote, "I would like to study computer science, but it may be hard to get a job in Africa with that degree. So I will study economics or public administration at the University of Liberia when I graduate from high school." A bit flabbergasted by this misconception, I responded to his e-mail and explained to him that the advent of the digital economy has kindled an unprecedented demand for professionals in the fields of computing and ICT in Africa, especially in Liberia. I told him that Africa as a whole has delayed in adopting information and communications technology but has begun to make progress; hence there will be an increase in demand for ICT and computing professionals. Liberia, now recovering from a protracted period of civil war, has been given a clean slate and will require the building of a new infrastructure. This new infrastructure, I said, will need computing and ICT professionals to design, build, and maintain. Obviously, my e-mails began to resonate with my nephew, and we continued our discussion in real time during an online chat session. The impact that the chat had on me led to the writing of this chapter.

Although I did succeed in persuading him to pursue a career in the field of computing, I could not help but wonder about other students in Liberia who aspired to embark upon similar careers but who, like my nephew, may be misinformed. To be successful in convincing these students that there is hope for ICT professionals and careers in Africa, we must initiate a debate on the topic, especially in Liberia. In this way, we will create awareness and allow students to make intelligent and educated decisions in determining their path for the future. I believe that this debate will impact individuals with dreams of becoming ICT professionals but who also feel totally emasculated by Liberia's maintenance of the status quo.

There are a plethora of reasons why a person should consider a career in the computing or ICT field in Africa, especially in Liberia. I could give you an exponential number of reasons, but I'll limit myself to just a few.

The impact of the digital economy on Liberia—At present, Liberia is in its infancy with regard to technology and by virtue of this will provide excellent job prospects for future careers in ICT. Liberia's economic future is bright, and computing and ICT professionals will play a major role in rebuilding the country to a level of modernity that will make it the pride of Africa. Currently, there has been a blitzkrieg in the demand for ICT professionals in Liberia because of the proliferation of financial institutions, telecommunications companies, other businesses, NGOs and the government. As more companies come to trust the stability of Liberian democracy, there will be a rush to invest in the country. Schools, especially institutions of higher learning, that are responsible for producing Liberia's future leaders will need more professionals, not only to teach, but also to run programs, administer systems, and hopefully perform research that could lead to novel innovations. The prospect of getting a job in the fields of computing and ICT in Liberia is very promising.

Growing number of submarine fiber-optic cables for broadband connectivity in Africa—The advent of the submarine fiber-optic cable, through which Liberia will gain broadband connectivity, will ignite an influx of investors in the country that will lead to new business development and great demands for ICT and computing professionals.

Organizations or firms will have to adapt to technological changes or risk losing their business— Some years ago, before I briefly worked for the American Power Conversion Company (APC), I recall making an interesting remark during the initial interview for that job. I recall informing the interviewing committee that any company that rejects, avoids, or is slow to adapt to technological changes will be out of business in less than six months. Even though I got the job, I felt in some way that my vision had "cursed" several companies, because years after that interview, many companies began reengineering their processes, while those who didn't, had to close their doors. The point I am making is that computing and ICT professionals are and will be needed to align technologies with business strategies; hence they have now become indispensable and mundane … just like accountants. This demand in services that computing and ICT professionals enjoy should be a motivation for students to join their ranks.

Western countries' shift in paradigm to regain superiority in the world economy—Western countries, especially the United States, have lost most of their manufacturing jobs to developing countries. This migration is due in part to the industrial nature of the jobs and the fact that these countries have mastered this kind of work and are willing to do it for a lot less. A job making a product that can be made in Liberia (for example) for $1/hr will not be sent back to the United States where a company will have to pay the factory worker $12/hr. This is just basic economics! So for Western countries to regain their place in the manufacturing business, they will have to develop newer strategies and processes involving new or advanced technologies that will require qualified individuals with education in computing and ICT. President Barack Obama's strategy of investing in new technologies (green technology, health-information technology, etc.) is evidence that the future workforce will gravitate toward computing and ICT. With globalization, any new paradigm adopted by Western countries will spread faster to other parts of the world; hence the promise of a great career and future in computing and ICT.

The global availability of computing and ICT jobs—A degree in computing or ICT provides an excellent opportunity for a person to get a job anywhere there is technology. Firms with information systems will need to provide information to their stakeholders, and this is best done by using information systems and technologies. Business processes will need to be run on information technologies, which will require ICT professionals for design and implementation.

Very rewarding salaries—Salary levels in computing and ICT professions are very rewarding and will continue to be as we move ahead. New innovations will bring a multitude of new jobs with better salaries and standards of living.

Satisfaction as a result of your impact on the people—A career in computing or ICT can be very gratifying and enjoyable. Think about the impact you will have on people and nations. I am sure Bill Gates (Microsoft), Linus Torvolds (Linux), or Jerry Yang (Yahoo!) never thought about the impact they were going have on the world. Today, they are not only rich and famous, but they are global agents of change. Anyone with vision and determination can do the same.

There are many more reasons that I could list as motivation to choose a career in ICT, but every individual will have a different and personal reason that will lead to a particular choice.

Different Categories Of Jobs In Computing

In the computing and ICT industries, there are a variety of jobs. This is because different companies have different needs for their information systems and business processes. Diverse systems require a variety of skills to be run properly and efficiently. I will categorize the types of skills firms seek from computing and ICT professionals by their functions: there are companies that use ICT with their core business strategies; there are firms that make computers; companies that make products that incorporate computers (cars with embedded systems); and there are companies that produce and supply software (Microsoft, Oracle, Adobe, etc). These categories produce a wide variety of jobs in the computing industry. Every company has computers in the workplace, but different kinds of companies will need different types of skills.

The list below illustrates a few of the jobs found in the computing and ICT areas. The list is not exhaustive but is long enough to give a student who aspires to embark on a journey in computing or ICT a reasonable idea of the breadth of options that are available.

- Computer scientist
- Computer engineer
- Information systems manager
- Information technology manager
- Computer programmer
- Data analyst/programmer
- Systems administrator
- Systems analyst
- Hardware engineer

- Software engineer
- Telecommunications/network specialist
- Web developer
- Embedded systems analyst/designer
- LAN administrator
- Computer salesperson
- Customer support/technical support
- Project manager
- Academic researcher
- Consultancy
- User training/technical support
- Security specialist

Listed above are careers that exist currently but did not exist twenty years ago. And just as we did not know they would emerge twenty years later, so it goes for the next twenty years.

Categories Of Studies For Computing And ICT Jobs

Computer science—Computer or computing sciences deal with the study of the development of the physical and theoretical design and understanding of computers. Students study hardware, software, network operations, digital design, software-development methods, engineering software, discrete mathematics, digital logic design, computer architecture, operating systems, analysis of algorithms, usability engineering, artificial intelligence (AI), etc. A bachelor's degree in this area prepares a student for a technical position in a firm. A master's or PhD degree prepares a student to become a decision maker, consultant, advisor, leader, or researcher.

Software Engineering —Software engineering is the application of engineering to software. It integrates mathematics, computer science, and some practices of engineering. Students graduating from this area of study usually work for companies that design and market the software (Adobe, Microsoft, etc.) that is used with computers. Personally, I question the appropriateness of referring to software engineering as an "engineering" subject. But that's a topic for a different day!

Internet Computing /Internet Engineering/E-Commerce—These areas of computing usually focus on the kind of technologies and skills needed to develop Internet-based systems. These areas include but are not limited to web technologies, network/Internetworking technologies, and databases (backend side). Internet engineering involves the underlying theory of Internet computing. Internet engineers usually work for companies like Cisco Systems or large Internet services providers. Internet computing focuses on the underlying theory of Internet computing and e-commerce technologies (the information and communications infrastructure) as well, although it includes studies in applications that run the Internet. E-commerce is basically commerce on the Internet. It focuses on the business and financial implications of the Internet. It emphasizes the underlying information and communications infrastructure. Computing students who take the e-commerce route always take courses in business and accounting.

Information Technology /Business Information Technology / Information Systems—The information technology area deals with the design, development, installation, and implementation of computer systems and applications. It takes a more technical approach toward computing; hence, students who take this route are usually real-world problem solvers using information technology solutions. Individuals who earn a bachelor's degree in this area are prepared for technical jobs in firms, while those who earn masters or doctorate degrees become ICT consultants, ICT leaders, etc. Business information technology provides computing and ICT skills as well as quantitative and modeling techniques to develop and implement sophisticated business-related computer systems. Information systems take a more businesslike approach to computing and involve the implementation of effective solutions to meet organizational and management needs for information and decision support. Students who gravitate toward this area mostly focus on the business aspects of computing.

I often encourage students to get at least a bachelor's degree in computing or ICT because it gives them competitive advantage over others who may be seeking similar positions and prepares them for leadership positions within firms. Alternatively, I advise students to seek certification programs that can give them the hard and soft skills needed to succeed in the workplace. Unfortunately, Liberia does not have a lot of institutions of higher learning that provide degree programs in computing or ICT. The next few years will see more ICT schools and perhaps certification centers in Liberia. Vocational institutions that provide computing and ICT certificate programs have become more available and provide a path to a successful career in computing and ICT. Students in the mathematics, engineering, science, and business programs at universities can still graduate with those degrees and work in a computing or ICT firm provided they supplement those skills with basic knowledge of current trends in computing and ICT.

In light of the predicted changes poised to occur in the area of ICT in Liberia, I encourage more students to engage in the fields of computing and ICT. Liberia has a bright future that will be significantly impacted by modern technology. The country will need those of us who have already garnered education in these fields, as well as the future leaders of Liberia who are aspiring to join in the building of a new and digital Liberia.

Chapter 11

Liberian Women Must Be Encouraged To Pursue Information And Communications Technology (ICT) Careers

When we formed the Students Working to Advance Technology (SWAT) team at B. W. Harris School, I emphasized the team's need for diversity in gender and grade levels. The goal was to expose and encourage young Liberian females to explore other industries of work that are typically dominated by men. While I managed to achieve my goal of balancing the SWAT team, gender imbalance in information and communications technology (ICT) and in many other areas in the work environment remains an issue, not only locally, but globally as well. This chapter will discuss gender disparities in the Liberian ICT community and its role in the digital economy. The role that women will play in Liberia's economic recovery will be critical as they will transcend the traditional roles filled in the past.

In Liberia, the relative balance of gender in government and other sectors of the society quintessentially indicate that we have the capacity to create balance in the area of ICT as well. Creating this balance, of course, requires role modeling and an overwhelming support on the part of the Liberian authorities. The women in Liberia need to be encouraged, motivated, determined, and willing to explore new areas of employment other than the traditional occupations/careers with which they and society are comfortable.

The problem I see with gender disparity in ICT is that Liberian women are not encouraged or motivated to embark upon careers in ICT or in the sciences. They ignore careers in ICT because there is a lack of role models with whom they can identify. Also, there are persistent stereotypical views that ICT is better suited to men; then there is a knowledge deficit regarding the many types of available ICT jobs. Interestingly, females generally enjoy using ICT and more often become competent users of computers and the Internet. An example of this is the frequency and proficiency with which many Liberian girls use social networks like Facebook and Hi-5. Moreover, a huge number of females in Liberia have a Yahoo! e-mail account that they use proficiently to communicate with relatives and friends locally and abroad.

Undoubtedly there has been much progress in the area of gender equality in the global workplace. Even Liberia has made significant progress in gender balance, but there is still a lot of work to be done in this area. The Ministry of Gender, which is the government's engine for promoting gender equality, is doing a great job in encouraging gender balance in many industries, but there is still more it can do in the area of ICT. The ministry will need to initiate programs and incentives that will empower and encourage women to engage in ICT as a way of supporting the Poverty Reduction Strategy. Policies that favor girls must be put in place to ensure that Liberia achieves a high percentage of ICT manpower ten years from now. Failure to initiate capacity-building programs that include females will result in the women of Liberia being further marginalized from the global community. This is why all stakeholders have to be fully committed to this initiative. There is hope, of course, as the country's national ICT policy document addresses the issue of women's role and access to ICT. The document acknowledges the marginalization of women in ICT and illustrates the government's willingness to address that issue (Ministry of Post and Telecommunications 2009).

In December of 2008 during our "hands-on" computer hardware training at B. W. Harris School, one of the female students grabbed a screwdriver and began to disassemble a computer as I had previously instructed her to do. As soon as she began to unscrew the screw that held the cover in place to her assigned computer, a male student yelled, "What are you doing? That's not a woman's job; women are only supposed to type letters on computers--not to open them." Ignoring those comments, the female student proceeded to open the computer without hesitation. Before the male student could say another word, I interrupted and gestured him toward a photo that I had accessed online through my Palm Pilot; it was a photo of Barbara Desoer, global technology and operations executive at Bank of America. Then I mentioned the name of Heather Jackson, chief information officer of Halifax Bank of Scotland. But before I went further to mention another foreign name, I mentioned the name of Ciata Victor, who is the Webmaster of TLCAfrica.com and a Liberian female with quite an IT background. I informed him that ICT is not a gender-specific discipline and that I had two females as my workmates and they were the smartest programmers that I had ever worked with. I also told him that there are a lot of Liberian female ICT professionals in the United States and around the world who, upon return to Liberia, will play a significant role in the rebuilding process.

The absence of Liberian women in ICT represents a double loss; not only for the ICT sector, which already is faced with a shortage of skills, but for women themselves, who miss further opportunities to enter the labor market. This gender gap will not be closed until Liberian stakeholders do more to educate, support, and encourage young women to explore careers in ICT. This means that public-private partnerships and collaboration will play a major role in changing perceptions about industry by giving access to more realistic and authentic information about ICT and its associated careers. There will be a need for closer cooperation between institutions of learning and the Ministry of Education, those of us in the Diaspora, and NGOs to encourage ICT awareness among young female students in Liberia.

Liberian schools will also have to play a major role in this effort. They will have to integrate ICT in their curricula, as is indicated in the national ICT policy paper, and include girls in all ICT-related activities to encourage them to pursue studies in ICT. They should also set up field trips and allow female students to visit NGOs, public and private entities, USAID, and any entity that shows females working in areas of ICT. This will encourage young girls to aspire to have careers in this field versus the traditional careers they enter because they feel they have limited choices.

Finally, the fourteen-year civil war set Liberians back phenomenally, but as my father once told me, *"Every setback is a setup for a greater comeback!"* The setback brought about by the war has set the stage for a greater comeback through the current recovery process. We see this in the progress that is being made in the area of gender equality. Today we have a Ministry of Gender as well as other organizations that advocate gender equality. That being said, Liberian gender advocates will have to escalate their efforts to ensure gender equality in the area of ICT as well as leverage its transformative potential to benefit women. Women in Liberia will need to take the initiative of engaging in ICT so that they will be able to enjoy the benefits of modernization and economic development that will come. Liberia currently has a disproportionate number of males to females in ICT, which is a serious problem that must be addressed in order to achieve a balanced ICT workforce.

CHAPTER 12

INTEGRATING TECHNOLOGY IN SCHOOLS REQUIRES A NATIONAL EDUCATIONAL TECHNOLOGY PLAN

No society can succeed if its people lack good education. Education is the key to a successful society, one that aims for prosperity. Liberia is a country that has always had a high illiteracy rate. Efforts to bring a decline in the illiteracy rate endure, but those efforts remain frozen in time with traditional educational approaches. We are in the twenty-first century, and now is the time for a change; a change that reflects the dynamics of the digital community.

To accomplish this, there must be a paradigm shift in Liberia's educational system. This shift in paradigm must include technology and means. We must begin developing a national education technology plan (NETP) that requires the integration of technology in all schools in order to prepare the new generation of Liberians for the challenges of the twenty-first century. This plan should, among other things, require students to take technology courses starting from, at a minimum, the tenth grade and continuing upwards. This will allow students who cannot afford to go to college after graduation from high school to be prepared for entry-level jobs. In this chapter I discuss the need for a national educational technology plan that can be used as a framework for capacity building in Liberia.

Applying for a job these days requires at least two things: a high school diploma and knowledge of computers and their applications. This is becoming a global requirement, and it is time we act to ensure that Liberians do not miss any more global opportunities. There are many students in Liberia who have been turned down by these four words: *must be computer literate.* These four words are now becoming a requirement to get a job in Liberia where technology has a very low penetration. Students graduate from school with no knowledge of computers or their applications. Only private schools that can afford to run computer labs or a technology program have the capacity to graduate students who are at least prepared to enter the job market. While the clause *must be computer literate* has become common across the global spectrum, it has without a doubt become an "opportunity killer" for Liberian and probably many African or developing countries' students who graduate from high school with desires to enter the workforce as an alternative to college. In response to the lack of comprehensive computer/technology-educational programs in Liberia's schools, most of their graduates are constrained to attend computer schools after graduation—which they can barely afford—to obtain the requisite skills needed to apply for jobs.

Education is the key to Liberia's economic growth and prosperity and its ability to compete in the global economy. It is the conduit to good jobs and higher earning potential for Liberians. It fosters the cross-border and cross-cultural collaboration that are desperately needed to solve the most challenging and evolving problems of our time. But modern education requires the integration of technology. And technology as we all know is at the core of virtually every aspect of our daily lives and work; therefore, it is imperative that we leverage it to provide engaging and powerful learning experiences.

The national educational technology plan mentioned above will espouse a revolutionary transformation of the Liberian educational system. It will discuss innovation and prompt implementation, regular evaluation, and continuous improvement. The NETP will lead to a path where technology plays a key role. It will kindle a shift from the currently practiced twentieth-century educational paradigms toward twenty-first-century student-centric paradigms. Pedagogy will gravitate toward an approach in which teachers play the role of facilitators and not sole deliverers of information in the classroom. This will allow students to think critically and engage in the learning process more enthusiastically.

To begin this initiative, the government of Liberia through the Ministry of Education will need to establish a steering committee that will encompass the following: universities, representatives of primary and secondary schools, ICT professionals, educators, representatives from the Ministry of Education, and other stakeholders. This committee will be charged with the development of a document that will lead to a revolutionary transformation of the Liberian educational system. The committee will:

- Research the Liberian education system and its current approaches and determine how technology can be integrated into those approaches or how new approaches can be incorporated.
- Determine a national educational technology mission and vision.
- Determine the needs and desires of Liberian schools both public and private.
- Provide short-term and long-term goals and a plan to implement them. In addition, a budget for technology should be created for each goal.
- Explain where the funding for technology will come from, what to do if funding is no longer available, or if more funding is needed.
- Create steps to evaluate the technology plan on a regular basis and allow for changes based on new technology.

Currently, Liberian schools lack two main things that are required for twenty-first-century education: the technological infrastructure that enables this new pedagogical paradigm and teachers who have the training in and knowledge of this new approach. Most, if not all, Liberian educators lack experience using technology that is part of the daily lives of professionals in other sectors. The same can be said of many of the educational leaders and policy makers in schools and the higher-educational institutions that prepare new educators for the field. Therefore, professional development (pre-service and in-service) should be strongly encouraged and changed from the current format being used to one that leverages online communities for collaborative learning. Collaborating with educators in the global community will greatly expand the "horizons" of Liberian educators.

This NETP will be an opportunity for change and when implemented will revolutionize the Liberian educational system. This change will be driven by emerging technology and the national need to radically improve the country's educational system. The Ministry of Education has to play a major role in identifying effective strategies to ensure the implementation of this plan.

Finally, advances in ICT have given us a multitude of opportunities to make a difference in the world. If Liberia invests in the creation of a knowledge-based society, it will reap tremendous benefits. And in a knowledge-based Liberia, Liberians will be able to develop several "products"; not only rice, palm oil, or petroleum, but software. Liberia may not parallel India, but I am sure it has the potential to parallel countries in East Africa that are currently involved in designing computer software in their local languages using open source technologies. Liberia will be capable of competing with other African countries in intellectual-property production once its educational system is revamped and revolutionized.

Part III

Government

Chapter 13

Electronic Government (E-Government)

Electronic government, most commonly referred to as e-government is, very loosely, a marriage between government and information and communication technology (ICT). This chapter covers the use of ICTs in government and how it (ICT) can improve government relations and services with its citizens and other stakeholders. This chapter will also discuss e-government and its delivery models as well as how e-government can work in Liberia and the impact it will have on the country. Note that this is not an attempt to be or act as a government expert but rather an attempt to align ICT with government for the betterment of the Liberian society. Also, I am not advocating technology as a substitute for good governance or good policy making. What I do advocate is the use of technology to run an efficient government, gain public trust, and be able to respond to the needs of society in an efficient and timely manner.

Since we started using the word, e-government has been defined in several ways by various individuals and entities. The United Nations Public Administration Network (UNPAN) refers to it as "utilizing the Internet and the world-wide-web for delivering government information and services to citizens" (UNPAN 2008), while E-government for Development defines it as "the use of information and communication technologies (ICTs) to improve the activities of public sector organizations" (eGovernment for Development n.d.). The Center for Technology in Government at the University of Albany defines e-government as "the use of information technology to support government operations, engage citizens, and provide government services" (Government n.d.), while the Liberian national ICT policy draft defines e-government as "the delivery of government-related information and services to the public through the use of ICT" (Ministry of Post and Telecommunications 2009). This includes services through wide area networks, local area networks, kiosks, the Internet, and fixed and mobile networks (Telecommunications 2009). Personally, I prefer the University at Albany's (State University of New York) and the Liberian national ICT policy document's definition of e-government. I believe that e-government is a government's way of using ICT for efficiency and cost savings in their daily operations, ensuring accountability and transparency, providing services to its citizens—in both the private and public sectors—and aligning with other stakeholders, including NGOs, the international community, etc. However, all of the definitions given above share one thing in common: government's use of ICT to achieve a goal.

E-government incorporates four categories or dimensions: government-to-citizen, or G2C; government-to-government, or G2G; government-to-business, or G2B; and government-to-employees, or G2E. Below I briefly explain the four categories of e-government.

- ***Government-to-government*** —This is "the use of ICT to improve the management of government, streamlining business processes to maintaining electronic records and improving the flow and integration of information" (Government n.d.), among all sectors of government. It involves cutting process costs, managing process performance, and connecting various arms of government to enhance the government's investigative and developmental capacity in addition to its implementation of those policies that guide its operations.

- ***Government-to-citizen*** — This is the use of ICT to increase citizen participation in the public decision-making process, providing them with details of the public sector and government activities. It is an interactive medium or media that allows citizens to voice their views and take action, provides cost-effective services that will create convenience for them, and above all, makes public servants accountable to their employers—the Liberian people. In G2C, access is given to citizens through a single window; most often, a Web site or a portal. This is a one-stop shop for citizens to be involved with their government and in the running of their country.

- ***Government-to-business***—This is the interaction between government and businesses using ICT. This digital connection between businesses and government enhances procurement processes, revenue collection, and several other activities that improve quality, convenience, and cost. It also involves the electronic exchange of goods and services between government, businesses, and citizens; for example, citizens wanting to pay their electric bills to the Liberia Electricity Corporation, taxes to the Ministry of Finance, or vehicle registration.

- ***Government-to-Employees***—This is the interaction between government and its employees through ICT. It involves a government's ability to empower its employees, enabling them to provide better support to the citizens of the country. What I like about this and what I feel will do Liberia a lot of good is how it streamlines internal processes and facilitates and improves collaboration and the sharing of knowledge. I am confident that with an effective "G2E" platform, taxpayers' money will be spent properly in Liberia.

E-government in Liberia has already begun, although it is still in its infancy. The executive mansion's Web site as well as the Web sites of other ministries and autonomous agencies illustrates the genesis of what could be the most efficient, transparent, accountable, democratically charged government initiative in the history of Liberia and, most likely, the subregion of West Africa.

According to a United Nations annual e-government survey, which includes a section titled E-government Readiness, West Africa recorded the lowest regional index. Its score was recorded as 0.2110 (UNPAN 2008). Astonishingly, Liberia, which did not have an online presence in 2005, ranked 163 in that survey, which was very impressive considering the country's current position in the area of ICT.

The United Nation's annual e-government survey is a comparative ranking of the countries of the world subsuming two primary indicators: i) the state of e-government readiness; and ii) the extent of e-participation. This survey assesses all 191 member-states of the United Nations using a constructed model for measurement of digitized services. It (the survey) uses a quantitative composite index of e-government readiness on the basis of website assessment, telecommunications infrastructure, and human-resource endowment (Nations 2008).

While the government of Liberia's Poverty Reduction Strategy does not say much about e-government, it mentions several mandates that would ultimately lead to the creation of an e-government initiative (Republic of Liberia 2006). One of the mandates was the creation of the national ICT policy, which I briefly discuss in Chapter 17.

The current e-government initiative embarked upon by the Liberian government has not generated interest among Liberians, nor has it garnered the type of operational efficiency that is expected from an e-government installation. The lack of infrastructure and awareness and the fact that the government's e-government efforts are primarily static Web sites with oftentimes outdated information does little for citizens who actually need information to make decisions. This is contrary to the purpose of an e-government effort that aims to provide and improve public-sector services, information, transactions, and interaction with government.

Benefits

I know it may sound redundant to mention some of the benefits of e-government when I have already said much about it above. But permit me to provide a few more detailed benefits:

- Convenience and cost effectiveness for businesses and the public benefits obtained from e-government

- Current public information is made available to citizens without them having to spend time, energy, and money to obtain them.

- E-government simplifies processes and makes access to government information more easily accessible for public sector agencies and citizens (Nation 2008). For example, the Ministry of Commerce can simplify the process of registering a business by digitizing existing forms and the requirements needed to register a business; the Liberian National Police can certify driving records to be used in courts especially for situations that involve traffic violations that are contested. This can reduce cost and eradicate unnecessary delays.

- "E-government can facilitate communication and improve the coordination of authorities at different tiers of government within organizations and at the departmental level" (Nations 2008).

- "It can enhance the speed and efficiency of operations by streamlining processes, lowering costs, improving research capabilities, documentation, and record-keeping" (Nations 2008).

The benefits mentioned about are just a few that e-government provides. And Liberia is in desperate need for efficiency in public services, transparency, accountability, and cost reduction.

There are several West African countries that have been successful with e-government. Some of those countries worth emulating include the national portal of Burkina Faso (www.Primature.gov.bf) which according to the UN E-government Survey of 2008, "is the only African portal which allows for online consultation; the Ministry of Finance of Cape Verde (http://www.minfin.cv) which created a one-stop shop with downloadable financial forms and statistics, and provided access to the ministry's database and archived information; and the Ministry of Health of Senegal (www.sante.gouv.sn), which enhanced its Web site to interact with its citizens" (Nations 2008).

The next wave of progress that the Liberian government can embark upon is an e-government program that has interactive Web sites, thereby giving citizens the opportunity to participate in activities involving government. The government must begin to use Web 2.0 technologies to garner more citizen participation instead of using Web 1.0 technologies that only allow for the publishing of documents on the Internet. Also, when creating an e-government platform, Liberia would benefit from using a twenty-first-century approach involving collaboration among the people who will use the service. Involving the people in the creation of a "tool" that will help improve their way of living is, in my opinion, very democratic and progressive.

Finally, I cannot conclude this chapter without mentioning e-readiness. E-readiness requires that we have three main things in place before embarking on an e-government initiative. They include enabling policies, an ICT infrastructure, and human capacity. Overall, the success of an effective e-government initiative in Liberia begins with organizational and technological readiness. Also, governance has to parallel the "new tool" (e-government) that will be implemented. Liberians should not implement a new technological solution with expectations of fixing the current system when the method of governing remains the same. Adapting to the new paradigm may be challenging, but ultimately goals will be achieved.

Liberians must embrace innovation and be innovators or we run the risk of remaining in the league of the poorest and most underdeveloped countries. With the right e-government infrastructure, improved technology—Web and otherwise—and trained human capital, Liberia will be able to transform a government system that is inefficient, unaccountable, inconvenient, and lacking transparency to one that is efficient, accountable, cost-effective, and transparent.

CHAPTER 14

USING THE WEB FOR TRANSPARENCY AND ACCOUNTABILITY: EMULATING OBAMA'S RECOVERY.GOV APPROACH

Building transparent systems as mentioned in the previous chapter can improve governance. Several technology solutions can be used to accomplish this initiative. This chapter takes a look at the use of the Web by the Liberian government to ensure transparency. It subsumes the U.S. president Barack Obama's "Recovery.Gov" approach, an initiative aimed at engaging the American public in American government activities, which hopes to ensure transparency and accountability. The use of Web technologies to ensure transparency has become common and part of modern governance. Liberia can surely use an innovative approach of this type to accomplish the same goal. I believe that the recommendations made in this chapter can be used by the Liberian government to build a more transparent system for accountability and transparency.

Liberia is a country that is said to be rich in natural resources. Yet, according to Wikipedia, it remains one of the poorest nations in the world, with 85 percent unemployment, primarily relying on foreign aid and humanitarian assistance (Wikipedia, Liberia—Economy n.d.). One of the causes to which Liberia's high poverty rate is attributed is corruption, which also plagues many other African nations. Some of those attempts have involved coup d'etats and other civil uprisings.

But those sometimes violent attempts to eliminate corruption often led to the installation of yet another corrupt regime. Liberia has had its share of attempts to eliminate corruption, but all of those attempts have either failed miserably or exacerbated the situation, leading to even higher levels of poverty and greater suffering on the part of the Liberian people.

When President Ellen Johnson Sirleaf and her administration took office in 2006, it was expected to be a government that would make a change. Those who constituted the government were trained professional Liberians who lived and worked in the Diaspora, and many thought they would bring and practice the principles of modern democracy. This brought hope to Liberians. The perception that this new group of professionals would run a government free of corruption and similar to those governments of developed countries from which they came brought hope to a totally emasculated Liberia. But Liberians were later disappointed when they discovered that many of the new government officials who claimed to have returned to rebuild the country were actually back home to personally aggrandize themselves by siphoning government funds. Corruption became very conspicuous and rampant. And when it seemed these acts of corruption were being perpetrated with impunity, the Liberian press began to weigh in and take a more investigative approach, which led to several revelations. These revelations have been very instrumental in influencing the national consensus to fight corruption.

In June of 2010, upon her return from the United States, President Ellen Johnson Sirleaf called on her cabinet to take stronger measures aimed at stamping out corruption (Kpayili 2010). The president, it seemed, had embarked upon a campaign to aggressively fight and defeat corruption. I thought that was a highly commendable initiative considering the impact that it (corruption) could have on her government as well her legacy. While I embraced her efforts to fight corruption, I thought the president should have advocated a more modernized, aggressive, and robust approach to fight corruption. I believed that there was a need to shift from the old approaches to newer ones; approaches that would include all stakeholders so that they could be involved in government activities.

By a modernized approach I refer to twenty-first-century anti-corruption mechanisms that entail the incorporation of ICT. ICT can be used to create an interactive medium (Web portal or social network–like medium) that requires officials entrusted with taxpayers' money to publicly disclose how funds are allocated, thus ensuring transparency and lessening corruption. This is currently the practice of the U.S. president Barack Obama's administration and can be experienced on the Web site www.Recovery.gov, which is a substantive test of their ability to be transparent in the use of technology.

What is a transparent government? A transparent government is one in which its citizens are able to actively engage in government activities and are apprised of the public policies and actions of public officials. Any government that is not transparent is susceptible to corruption and undue influence because it lacks public oversight of decision making. In twenty-first-century governance, the public has access to government activities via information and communication technologies (ICT)—using social networks, virtual worlds, video-sharing sites, etc. Take the Obama administration's attempt to ensure transparency by using YouTube, Facebook, and other social networks to provide public information about government activities. Effective use of Web 2.0 technologies in governance can be very beneficial to both the government and its citizens. Governments alone do not have all the answers to run a nation; they need the expertise of the public. Therefore engaging the public through interactive media, over the Internet, is fast becoming commonplace.

Liberia needs a publicly interactive medium that can promote transparency and accountability and provide information for citizens about what their government is doing. Information maintained by the Liberian government or any government for that matter, is a national asset and must be made available to the public. Also, to curb corruption, the government of Liberia will have to take appropriate action, consistent with Liberian laws and policies, to make information rapidly available and to solicit feedback from the public to aid in information gathering and proper governance.

Since the establishment of the General Auditing Commission (GAC) and the freedom being enjoyed by the press in Liberia, several stories of corruption have been published; prominent individuals have been exposed, re-situated, forced to resign, or dismissed. But even with knowledge of this, corruption is still rampant and more discoveries are still being made and published. This is extremely discouraging and undermines the good work that the government has done and continues to do.

The government of Liberia must encourage the public's engagement, because it enhances the effectiveness and improves the quality of its decisions. As I mentioned above, governments do not have all the answers; hence, they rely on the public, which has an amalgamation of knowledge from which public officials can benefit.

The government of Liberia will need to establish open media for transparency, perhaps through a mandate to *all* government agencies to utilize innovative technologies in order to make their information available to the public. This will also call for a collaborative approach with significant support from those in leadership positions. The ultimate product will be a merger of useful information stored in one location that provides a one-stop online medium for monitoring government activities.

How Can We Do This?

Initially, achieving a fully transparent government may prove challenging in a country like Liberia because of the proliferation and common practice of "under-the-table" transactions. But with much work and a good vision, achieving this goal is not only realistic but fairly easy. There are several ways to implement a transparent system using ICT. Below I list four of them.

1. Automate all processes that can be automated.

2. All government data should be made available to the public online (Web site, telephone help lines, etc.) and in real time.

3. Institute a social network–type of medium that encourages information sharing, collaboration, and interactivity. This medium should allow the citizens to discuss government activities based on documents that are made available online by the government.

4. Encourage feedback from the public and make sure action is taken where and whenever it needs to be taken.

The above mentioned approaches to achieving a transparent system in government can be done through a good e-government initiative. An e-government system that links all government ministries and public corporations should be created. All counties and local government activities within Liberia should be linked as well. An e-billing and e-procurement system should be established to allow the government to pay bills and obtain products efficiently while ensuring accountability and transparency. These systems would promote electronic transactions; payments made and received would be routed directly to designated banks and proper documentations made available online in real time. These measures would lead to a more efficient, accountable, and transparent government; one that the people of Liberia and the world could trust.

Most cynics argue that online governmental transparency will only be as good as its user; in other words, they believe that the information that would be posted on the e-government Web site could be manipulated to hide actual transactions. But in this information age, when the press has on-demand access to information through the use of technology, it would be impolitic for any government official to attempt to massage the truth.

For transparency to be ubiquitous, all stakeholders in Liberia must participate. Businesses will have to be transparent as well, and perhaps we might need an act that parallels the U.S. Sarbanes-Oxley Act that requires better and transparent reporting on performance of public and private companies.

I know that transparency works. Let me tell you about my most recent experience with transparency. Currently (2010), I serve as the second vice national chairman of the B. W. Harris Alumni Association, USA. I am also pretty much the association's as well as my alma mater's (B. W. Harris School) technology consultant. One of the objectives of our administration was to be transparent and accountable. Not only did we seek to be transparent, but we wanted the general assembly to be engaged in all of our activities. Therefore we gravitated toward the use of technology to achieve our objectives. As a way of being transparent, we sent out electronic copies of financial statements to the general assembly, utilized e-mails to disseminate information, used our own social network, which we call MySparta (Web 2.0 technology), held teleconferences through Freeconferencecalls.com, and utilized many other innovative ways to ensure transparency. The result of this was increased participation on the part of the members and increased support to our alma mater. I discussed some of this support in chapter 8. Also, you can find information on the B. W. Harris Alumni Association, USA, at www.BWHarrisAlumniUSA.org.

As I conclude this chapter, I must reiterate that the time has come for the government and other supporting organizations to begin to rethink how technology can be leveraged and deployed to eliminate the cancer of corruption that continue to plague the Liberian society. Government should not allow its vision for transparency to be clouded by the fear of change, cynicism, or the unwillingness to engage in new and innovative methods to provide solutions to Liberia's problems. Corruption has been and is still one of those ills of the Liberian society that continue to strangle development even in times when desperate efforts are being made to bring recovery to the country. It is time that an aggressive war is waged on corruption using ICT as the primary weapon of choice. This approach will certainly alleviate or lessen, if not eradicate, corruption in Liberia, thereby allowing revenue to be equitably distributed amongst Liberians. This approach will institutionalize Liberians and allow them to adapt to an open system that encourages their involvement and ability to monitor their public servants.

Chapter 15

Combating the Epidemic Of Corruption Through Open And Automated Systems And Processes

Corruption in Africa is endemic and systemic, and societies have yet to find a panacea. It has been a problem for a long time; as evidenced by the story of Zacchaeus in the Bible. Unfortunately, corruption persists in societies like Liberia despite sincere efforts to eradicate it. In the past there were few resources available to fight and eliminate corruption; hence, it was often perpetrated with impunity in many countries. But that has changed! The advent of the "information superhighway" has brought several new paradigms and technologies that allow knowledge and information sharing that help to expose corruption. Open and automated systems through the implementation of e-government solutions have changed the dynamics of societies and have become tools that are being used to curtail corruption. This chapter is about integrating technology in the Liberian public sector to prevent, alleviate, or eradicate corruption. Focus is placed on corruption in government, because historically this is where corruption has been the most rampant. The chapter looks to a future Liberia where the use of automated and open systems will counter corruption and allow the government to garner much-needed revenue.

Corruption in Liberia is nothing new; it has existed for as long as I can remember and before that. In fact, it seems that the country thrives on corruption, because despite all efforts—including a military coup and civil war—to bring an end to corruption, it continues to prevail and the breadth to which it has prevailed has grown even wider.

Public outcry against corruption continues with minimal impact. These acts of corruption have overshadowed the good work and strides that the government has made, making the government appear to the world as the stereotypical African government that thrives on corruption and the exploitation of its people. The fact is that the current leadership (President Sirleaf) is known to be a strong opponent of corruption, but some the individuals with which she has entrusted the affairs of the country and brought to help in the country's recovery process have disappointed her. And in response to her disappointment, President Sirleaf decided to dismiss or ask several of those accused of corruption to resign their positions. But does this solve or prevent corruption in a country that is engulfed by it? Absolutely not! It only creates room for others to come in and perpetrate the same acts that were perpetrated by their predecessors.

So how can this problem of corruption that is so interwoven in the fabric of Liberia be solved? What can we do? The answer is in implementing new ways to do old things and above all else exploring the unfamiliar to locate new opportunities. And this is what a digital Liberia will do in terms of fighting corruption.

The solution that a digital Liberia provides for the fight against corruption is the use of open and automated systems, a radical shift from the status quo to novel approaches in every aspect of the society. This shift in government has already begun and is illustrated by the freedoms experienced by the press and "whistleblowers" that expose the ills of society without fear of retribution. In addition, the excellent investigative reporting—free from bias, I presume—and the availability of information garnered through the utilization of technology leaves me to surmise that the days of public officials perpetrating corruption in the dark are long gone. Unfortunately, the press will not single-handedly solve corruption problems in Liberia. The government itself will have to buttress its current efforts to fight corruption by eliminating the urge or motive through instituting automated systems and processes. The use of open and automated systems through ICT will reduce cost, create efficient and productive work environments, and most of all, alleviate, if not eliminate, corruption. Those whose jobs become automated will be trained to perform new and different tasks since each change in the system will necessitate that.

I am not touting automation through the use of ICT as being the ultimate panacea for the eradication of corruption, but it does offer the potential to alleviate the day-to-day corruption that persists in the government and in other areas. Furthermore, automated systems do provide reports, and reports produce paper trails that facilitate the work of the press, auditors, and the public, thereby making it easy and quick to expose corruption. This is the type of open system that will help in the fight against corruption.

The use of open systems fueled by Internet technologies for the exposure of corruption is becoming a ubiquitous effort since governments around the world have begun to engage Web technologies to ensure transparency. Social networks, blogs, e-mails, and Web sites are a few of the Web technologies that are being used to expose corruption in this information age. ICT professionals, government regulators, and members of the fourth estate will have to work in collaboration using these Web technologies for collaboration and information sharing to achieve a transparent system. This is not a task for one group, and most importantly, the task is an exceptionally difficult one.

In business we refer to "business process reengineering" as the process of redesigning business processes to improve performance. In Liberia's case, a "redesign of processes" using modern ICTs can lead to an open, one-stop service system that can be delivered in an automated fashion. This revamped or new system should bring efficiency, lowered cost, accountability, and most of all, transparency. I list below both the open and automated systems that can be utilized in Liberia.

Open Systems

An open system should entail the digitization of all documents, forms, and processes that previously involved interaction between citizens and public officials. The system should be designed to accommodate the user from the time he/she expresses the need for a service to the final processing stages. It will be available to the public on the Internet 24 hours a day, 7 days a week, and 365 days a year. For example, business registration, drivers' license renewals, or the clearing of goods from ports of entry, which are all nonconfidential processes, should and can be done using Web services and applications.

This "open system" is intended to minimize all occurrences of corruption, irregularities, and inconvenience related to the processing of civil applications. This system will provide a medium for Liberians living in Liberia and around the world to access civil applications/processes anytime from anywhere without personally visiting government offices.

Another approach that should be used in the "open system" involves the use of e-mail in *all* official government communications in order to establish a paper trail to facilitate audits and public inquiries. The government will set policies that will classify all e-mails as public documents to further enhance the open system. Also, when I say e-mail, I am referring to the official e-mail system of the Republic of Liberia and not free e-mails like Yahoo!, Gmail, or Hotmail. These free e-mails should *never* be used for official business.

Automated Systems

Automated systems eliminate face-to-face interactions and create a medium where a trail can be left and tracked. They also eliminate the need for unnecessary bureaucracy, another process that leads to corruption. Processes that require humongous paperwork and procedures provide opportunities for middle persons to engage in corrupt practices. Obtaining a passport from the Ministry of Foreign Affairs, registering a business at the Ministry of Commerce, or expediting shipping through Customs at the Freeport of Monrovia are a few of the processes that require automation, not only for transparency and accountability but for efficiency and convenience.

A Few Things To Consider

Like all other new and expected changes, open and automated systems will present some challenges, but those challenges will be overcome as the new government processes are embraced and accepted by Liberians. Initially, some Liberians will become intimidated by these new changes. They will fear that these new systems and processes will take their jobs away and hence detest them not only because the processes might take their jobs but because they eliminate the status quo from which they benefited. However, none of these challenges will reverse the newly implemented processes or systems installed by government in a digital Liberia.

Other challenges, apart from government employee's rejection of the new processes, are listed below.

- **Cost**—The cost of implementation might be an issue initially, but as the government and its citizens begin to realize the benefits or the returns on the investment, cost will no longer matter. Also, the use of free and open source software (FOSS) provides a cheaper alternative.

- **Infrastructure**—Infrastructure might be another issue and may present some challenges but not obstacles that cannot be circumvented. The connection to the ACE submarine fiber-optic cable project in 2012 will enhance the need for an infrastructure to be erected. Otherwise, wireless networking, Web services, cloud computing, and virtualization technologies can be used as alternatives, especially when extending to rural areas where there is no infrastructure or where there might be a delay to implement an infrastructure. Using satellite connectivity may possibly run these systems but quality and cost would be major challenges to address.

- **Capacity**—This means that capacity building is imperative. This also means that several ICT schools and programs need to be erected and initiated. Most of all, the nation's universities will need to include robust ICT programs in order to produce qualified workers for the digital environment. Also, government will have to incentivize Liberian professionals around the world to return to assist in the capacity-building process through training programs, like "train-the-trainer," that will significantly impact capacity in Liberia.

- **Maintenance and technical support**—This will not be an issue provided efforts to build capacity are made and more qualified Liberians are employed in the ICT sector.

With the availability of new technologies, implementing, monitoring, and supporting systems and processes should not be a problem.

Benefits of Automated and Open Systems

- Automated systems will eliminate face-to-face interaction, minimizing corruption and making Liberians working in the public sector function more efficiently.
- An open system will make available on a 24/7/365 basis the whole process of civil applications, maximizing transparency in government and providing convenience for Liberians.
- An open system will significantly reduce the probability of any corruption and other improprieties and also enhance the effectiveness of internal and external monitoring and auditing.
- This approach encourages the injection of technology into the Liberian society, giving Liberians full use of the Internet and its technologies for economic development.
- New jobs will be created.
- Investors will no longer need to bring in foreign workers to perform IT services.

The government of Liberia and its citizenry must illustrate the will and vision to eliminate corruption. There also needs to be a partnership between government, nongovernmental organizations, and the private sector as well as the citizens for this process to work. This is by no means the creation of another medium to add to the albatross currently shouldered by the private sector but rather an amalgamated effort to fight corruption. I remain hopeful about the prospects of using ICT services to fight and eliminate corruption. And as I warned above, those who benefit from the inefficiencies of the current corrupt system will provide strong opposition against the probability or possibility of ICT curtailing or minimizing corruption. However, any battle to prevent progress will not be successful; the alternative to open and automated systems and processes is the continuation of the traditional approaches that are currently used to combat corruption, which obviously has not been working. And if the traditional approaches are ever used as an alternative, it would merely lead to the perpetuation of rampant corruption, which confines the country to unprecedented poverty and illiteracy and thereby strangles any effort at economic recovery.

The elimination of corruption in Liberia will not be an easy task even though we espouse the use of open and automated systems as a possible alternative. There will be a need for Liberia's "brain gain" situated in the Diaspora, to return to Liberia and work with other Liberians using modern strategies and technologies to fight corruption, which has for so long imbrued the society.

By reengineering government processes, implementing the appropriate controls and audit, and instituting mechanisms that ensure transparency and accountability, Liberia will set the stage for an equitable distribution of the tremendous wealth with which it has been endowed. The elimination of corruption in the Liberian society will free up resources and make them (resources) available for investments. These investments will provide more jobs for Liberians, enhance the country's endeavor to create a knowledge-economy, and improve the lives of Liberians.

Chapter 16

The ASYCUDA Software Project Of The Ministry Of Finance—An Example Of Process Automation In The Liberian Government

As Liberia moves toward being a digital society, automation will undoubtedly become ubiquitous. As previously mentioned in chapter 15 and an article published in the *Liberian Daily Observer* in 2009, implementation of said technology and systems will decrease unnecessary human interaction, which often leads to corruption, in the government and in public and private sectors.

When the Ministry of Finance of Liberia announced the Automated Systems of Customs Data, ASYCUDA, I wrote an article explaining ASYCUDA to the Liberian public (Wilkins, *Liberian Daily Observer* 2009c). Soon after I wrote that article, I received several e-mails from readers questioning this project's ability to alleviate corruption. In my response, I explained that ASYCUDA may not be the "silver bullet" that eradicates corruption in Liberia, but it has the potential to alleviate and enhance the existing customs system, which was inefficient and open to corruption.

The government of Liberia through the Ministry of Finance and the African Development Bank had announced the signing of a contract for the purchase of the ASYCUDA software at a cost 1.5 million dollars

(Kennedy 2009). This was a crucial move on the part of the government because the software would bring a shift in operational paradigm at ports of entry in Liberia, which had been rife with corruption.

The purchase of the ASYCUDA was an indication that Liberia had embraced ICT as an enabler in its recovery process. Ostensibly, the use of ASYCUDA in the customs sector was desperately needed because of the inefficiency, inconveniences, and corruption importers experienced at the ports of entry in Liberia.

This chapter briefly explores ASYCUDA, what it is as well as its functional and technical advantages. A brief discussion will follow about the total cost of ownership (TCO), the installation process, and what the government ministries and agencies must consider in managing such a "radical" change. I apologize for the use of several technical terms in this chapter. I had hoped to avoid usage of ICT jargons and terms because the book is being written for everyone to be able to read. I beg your indulgence and ask that you refer to the glossary if needed because a discussion on this topic cannot be had without the use of ICT terms and acronyms.

The Automated Systems of Customs Data, or "ASYCUDA, is a computerized customs management system that covers most foreign trade procedures" (Andrews et al. 2007). It handles manifests and customs declarations and accounting procedures as well as transit and suspense procedures. ASYCUDA was developed in 1981 by UNCTAD (United Nations Conference on Trade and Development) in answer to a request from the Economic Community of West African States (ECOWAS) to assist in the compilation of foreign trade statistics in their member states (Andrews et al. 2007). ASYCUDA is currently being used in more than eighty-three countries. It was developed in Geneva, Switzerland, by UNCTAD, and it follows international codes and standards developed by the ISO (International Organization for Standardization), the WCO (World Customs Organization), and the United Nations (UNCTAD n.d.). The software operates in a client/server environment under UNIX and DOS operating systems and RDBMS (relational-database management systems). Since it was developed, ASYCUDA has been through several versions: ASYCUDA Ver. 1(1981–1984), ASYCUDA Ver. 2 (1985–1995), and ASYCUDA Ver. 3 or ASYCUDA++ (1992–present). The most current version is the ASYCUDAWorld (UNCTAD, ASYCUDA: Automated Systems for Customs Data-Technology n.d.). To benefit from the latest innovations in information technology, the ASYCUDA Technical Development and Implementation (ATDI) designed and developed ASYCDA++. ASYCUDA++ uses object-oriented technology in a client/ server environment (UNCTAD, ASYCUDA: Automated Systems for Customs Data-Technology n.d.). This means that ASYCUDA++ contains the flexibility and portability needed to interoperate with a wide range of both Intel and RISC-based (reduced-instruction-set computers) platforms.

In March of 2002, UNCTAD officially launched ASYCUDAWorld. ASYCUDAWorld uses 100 percent java-enabled operating systems and independent servers (UNIX, Windows, Apple), and it is database independent (Oracle, Informix, Sybase, DB2, SQL server). It also runs on portable devices (PCs, PDAs, mobile phones, tablet PCs); ultra-thin (for mobile devices), thin (with major Web browsers), and thick clients (standalone clients) for Linux, Windows, and Apple-based computers (UNCTAD n.d.). As far as security is concerned, users' access is restricted to the functions of a group's profile. It also has auditing capabilities and asymmetrical encryption as well as built-in security features including electronic signatures and several levels and types of encryption algorithms. ASYCUDA has a long list of features that I am sure the Bureau of Customs will leverage in time to come. But overall, the software is carefully written and has been working well for countries in which it has been installed.

The installation of ASYCUDA is based on the request of governments, especially governments of developing countries. This is also done in collaboration with UNCTAD and its experts. The process begins with an initial assessment study done by a team of experts from UNCTAD. This assessment subsumes the conditions of the country, including the political situation and the country's human resources. Upon doing the assessment, the team reaches an agreement with the government regarding installation. Funding for the project is obtained and provided by the requesting country. In the event that there's a need for supplemental funding, UNCTAD does the negotiations with other funding organizations (UNCTAD n.d.).

The $1.5 million that was paid by the government of Liberia and the African Development Bank seems lower than what other countries have paid for the installation of ASYCUDA. I am not sure whether that amount at that time subsumed total cost of ownership (TCO). ASYCUDA should be a good tool for the Bureau of Customs and a great benefit to the people of Liberia in terms of automated processes, reduced corruption, and transparency.

Advantages Of Implementing ASYCUDA

Below, I list a few of the functional and technical advantages of the ASYCUDA software according to UNCTAD (UNCTAD n.d.).

Functional Advantages

1. Automatic calculation of custom duties/taxes
2. Automation of customs procedures, including but not limited to cargo control, in-transit operations monitoring, accounting, risk management, etc.
3. Document changes, tracking, and audit history
4. Use of UN, ISO, and WTO standards as well as WCO data model sets
5. Complete customs workflow design to automate the declaration-processing path
6. Allows digital images to be attached to customs declarations; images include photos of driver, truck, registration numbers, scanned documents, etc.
7. Built-in capability to support specific national requirements and frequent changes in an integrated system environment

TECHNICAL ADVANTAGES

1. ASYCUDA is independent of operating systems and RDBMS (relational-database management systems), thereby making possible implementation on a wide range of software and hardware platform including personal computers, portable equipment (PDAs, mobile phones, tablet PCs, etc.).
2. Possible inclusion of several types of clients, including thick, thin, and ultra-thin
3. Multi-language/alphabet (user-interface and data), Unicode E-documents
4. Robust built-in security features, including user authentication (group, name, and password), asymmetric encryption, PKI, electronic signature, biometrics, etc. Changes and updates of reference data completed without programming
5. Communications is done via Web, Internet, and Intranet. It includes full independence of telecom infrastructures and is very resilient to telecom breakdowns.

TOTAL COST OF OWNERSHIP

As I said above, I am not sure that the $1.5 million invested in the Liberia ASYCUDA project included total cost of ownership (TCO). From what I gathered, the amount was just the cost of implementation. My source was not certain and did not want to go on record. But what good is an IT implementation if total cost of ownership is not considered? TCO is a great analysis resource and an easy way to understand the true costs of an IT investment. "It offers a statement on the financial impact of deploying an information technology product over its life cycle" (Wikipedia, Total Cost of Ownership [TCO] 2010). In IT projects, software, hardware, training, and maintenance are considered elements of total cost of ownership; and their costs should be subsumed when planning to implement an IT project (Wikipedia, Total Cost of Ownership [TCO] 2010).

The total cost of ownership is a crucial aspect of an IT project that has to be thoroughly considered. The deployment of a new technology is more than just doing the installation. It is safe to say that the ASYCUDA project will ultimately cost more than $1.5 million. And since the intended platform of use will include the Microsoft Windows operating system, which is expensive but is the standard client operating system, the costs will undoubtedly be more than $1.5 million. Had the Linux operating system been used on the client side, the cost might be minimal because it is free of charge.

What Needs To Be Done To Implement Such A System At Other Government Agencies?

Moving from a partially manual environment to an automated one is a big change if not managed properly. ASYCUDA brought fear to a lot of employees who worked at the Bureau of Customs of the Republic of Liberia. Many feared that their lack of computer skills would cost them their jobs. This kind of fear is normal. Rarely are individuals initially comfortable with change; and thus their resistance! But when managed properly, change can be done seamlessly.

The best way to implement this kind of change in a workplace that faces challenges in using ICT is to create an extensive awareness program to educate those involved in the utilization of the new system. This must be done specifically with those having direct interaction with the software. This will reduce fear and concerns about the goals of the newly implemented system. Also, those authorities involved in this process must be committed and supportive in the implementation of the system and the training of their staff. Furthermore, stakeholders should not view change as the government's approach to phase out jobs but rather a means of bringing new developments, making systems efficient, creating more job opportunities as revenue is collected, and preparing Liberia for competition in the new and digital global economy.

Managers, on the other hand, should be creative and dynamic by routing all employees whose tasks will be automated to other areas or retrain them to become more effective and efficient in the performance of new responsibilities. Instead of downsizing or "rightsizing," managers should retain employees and eliminate those expenditures that are dispensable such as unnecessary trips, expensive cars, spending wastefully on social occasions, etc.

I have to applaud those individuals who came up with the decision to implement ASYCUDA in Liberia. That project was an illustration of the good vision of the authorities at the Ministry of Finance. Changing a previously "antiquated" system to an "automated" one is a twenty-first-century paradigm that can help reduce corruption. The road ahead for Liberia is very bright and has taken an irreversible path toward development. Initiatives like the ASYCUDA project must continue; otherwise we will miss the opportunity to be a developed country. Other government ministries and agencies should follow the Ministry of Finance's lead in the implementation of automated systems like ASYCUDA.

CHAPTER 17

THE NATIONAL ICT AND TELECOMMUNICATIONS POLICY DOCUMENT AND ITS IMPACT ON LIBERIA

The first time I reviewed Liberia's "National ICT and Telecommunications Policy," I became excited and hopeful of Liberia's future. It (National ICT and Telecommunications Policy) was made available to the public in 2009 on the Ministry of Post and Telecommunications' Web site; at that time it was only a draft. The document addressed several crucial issues concerning the future of ICT in Liberia. In my opinion, the basic premise of the policy illustrates that Liberia's economic and societal recovery and reconstruction can be achieved through the development, deployment, and utilization of ICT. This policy will serve as Liberia's road map or guide to the "new" (digital) economy.

In this chapter, I attempt to briefly analyze the National ICT and Telecommunications Policy and its potential impact on the country's future. Before moving forward, I must reveal my involvement with the development of the policy; I had the honor of making a nominal contribution in terms of suggestions and ideas.

But I must admit that I was quite impressed with the method used by the Ministry of Post and Telecommunications in making the document publicly available and requesting input from both local and international ICT professionals. Such willingness to garner a cornucopia of participation allows access to the collective genius of nations. This method of collaboration should be emulated by others to solve difficult problems. I submit that mass collaboration, especially in ICT has given rise to innovations that have transformed societies, if not the entire world. The Linux operating system, MySpace, Flickr, YouTube, etc., are all a result of mass collaboration. It (mass collaboration) is becoming a common and ubiquitous approach to problem solving and will do wonders for Liberia if this approach is adopted.

As I mentioned earlier, the National ICT and Telecommunications Policy Draft addresses several crucial national issues that will transform Liberia from a country known for poverty and illiteracy to a nation fully connected to the global community, knowledgeable, and economically viable. But for this to happen, this policy requires the full support of *all* stakeholders, including government, public and private sectors, NGOs, our international partners, and every Liberian.

The National ICT and Telecommunications Policy is now the Liberian government's policy for the ICT and telecommunications sectors. It was created to solve the problems that exist within the country's telecommunications sector. It espouses the development of a comprehensive ICT and telecommunications strategy that will bring growth and development to the country through support of the government's Poverty Reduction Strategy (PRS). Most of all, it will give the country the vision and strategy needed to create a digital Liberia prepared to engage the global community.

The National ICT and Telecommunications Policy mentions its responsibility to ensure that services and systems are people-centric, universally accessible, and cost effective. It delineates the Government of Liberia's role in establishing "legal regulatory frameworks and institutional mechanisms to guide the activities of all stakeholders." It aims to achieve "modernization and rapid expansion of the telecommunications network and communication systems" (Ministry of Post and Telecommunications 2009). The bottom line is the creation of a digital Liberia ready to increase productivity, share information, and engage the global community.

One thing that bothers me about this policy is that it sets an objective for five years, beginning in 2009 and ending in 2014. It is expected to align with the World Summit on Information Society's

(WSIS) projection that aims to connect the world by 2015. But the current pace at which ICT is progressing in Liberia does raise more questions about whether any of the goals, visions, and objectives listed in the policy can be met within stipulated timeframes.

The National ICT and Telecommunications Policy addresses issues such as e-government, e-commerce, e-health, e-education, e-agriculture, and e-security, and it goes on to establish a national ICT governing board to "oversee, monitor and evaluate the operations and implementation of ICT objectives and programs." This governing board consists of the President of Liberia as national chairman, the Minister of Post and Telecommunications as national co-chairman, the Ministries of Information, Internal Affairs, Planning and Economic Affairs and the Liberia Telecommunications Authority as members (Ministry of Post and Telecommunications 2009). While I agree that there needs to be a governing board, I strongly believe that the implementation of ICT objectives should be left to an independent body that centralizes ICT expertise. The governing board, as the policy proposes, may be overwhelmed with other responsibilities, thus resulting in an inefficient and ineffective ICT sector.

What I found most interesting in the "policy document" is the reference to the use of open source software for the development of applications that can be used for commercial purposes. I strongly support this goal and believe that the use of open source software will lead to the genesis of Liberia's entry into the software development market.

The policy brings hope to the student of the University of Liberia who is required to repeat several courses in another country because his/her credits are not accepted; it brings hope to the pregnant woman who has to travel to Monrovia for healthcare because her community lacks medical resources; and it brings hope to the job seeker who cannot obtain employment because he/she lacks the required computer skills. It also brings hope to a nation that is held captive by poverty, illiteracy, disease, and economic strangulation. Most importantly, the policy brings hope to a new generation of Liberians who will be responsible for leading the country into the next century. But this policy will be nothing but a document if it is not fully and unconditionally supported and implemented by the government and other stakeholders.

Louis Pasteur, a great microbiologist, once said that "chance favors the prepared mind." This is true in Liberia's case. There are a lot of "prepared minds" in Liberia in search of knowledge that will help bring development to the country. But the lack of resources has been an obstacle for many years. The injection of ICT into the Liberian society will give these "prepared" minds choices.

During one of my trips to Liberia, a seventeen-year-old student named Tarweh who lives around the Old Matadi Estate area sat by me at a local Internet café and curiously watched while my laptop booted (started). The look on his face indicated that he had either seen a ghost or was overwhelmed with curiosity. I knew he had some questions to ask because I was booting my laptop into Linux instead of Windows XP (I always carry two operating systems on my laptop). To him that was very unusual. He had not seen a laptop boot to any other operating system except a Microsoft Windows operating system. As the boot process progressed, he asked, "What's an algorithm?" I looked at him, astonished, before answering his question. I was not surprised that he stuttered while saying the word, but I was surprised at his seemingly fervent desire for knowledge. I felt that, with such an intellectual curiosity at his age, he could be quite an asset if he had the opportunities to learn as kids in other countries do. I thought that if he was taught the basics of computer programming, he just might be able to write his own computer applications! Tarweh is the quintessential example of Liberia's "untapped natural resource—the prepared mind" which the National ICT and Telecommunications Policy is expected to nurture for economic development.

Liberians should not allow the fact that electricity and drinking water have not reached rural areas to preclude them from bringing the twenty-first-century changes in areas that have access to those facilities. If Monrovia is the only place that can facilitate the implementation then by all means we should gravitate toward Monrovia. Development should occur concurrently. In other words, as electricity and water are established in rural areas, those areas that have had access to these amenities should be primed for ICT infusion. I think that it is better to have ten percent educated Liberians prepared and ready to take Liberia into the global economy in the next two years, than to have 100 percent of the nation wait for the next twenty years. By then Liberia will be farther behind than we are now. The ten percent mentioned above would at least have the capacity to educate the rest of the country that had been less privileged or marginalized.

National development will occur, although not all in the six-year term of President Ellen Johnson Sirleaf. But as Liberia gradually embarks upon national development, it must ensure that its national reconstruction initiatives parallel those of today's global setting and paradigms so as to avoid rebuilding what existed twenty years ago. In a nutshell and to repeat what I said earlier, every reconstruction endeavor in Liberia must be done in preparation for a digital Liberia.

The National ICT and Telecommunications Policy aims to bring a lot of changes to Liberia; changes that will develop Liberia and enable it to compete globally. Liberians *must* support and join those involved in this process to ensure that every aspect of that policy is fully implemented. Those involved in the process must continue to look at the "big picture" when it comes time for implementation because of the overall benefits to every Liberian.

Finally, Liberia faces several challenges in implementing this policy, and the process will be gradual. Issues surrounding infrastructure, capacity, and economics still exist and will go away gradually. Liberia is blessed with a multitude of resources that can be used to find solutions to the challenges it faces. Just think about the cellular phone revolution in Liberia; something the country never envisioned! What Liberia needs to do is declare independence from the captivity of fear, poverty, and illiteracy and move toward the sanctuary of hope and optimism. With hope and optimism, Liberians will be able to explore their creative genius and seek genuine and lasting economic solutions!

Chapter 18

Information And Communications Technologies for Development (ICT4D)

Information and communications technology for development (ICT4D) is often referred to when ICT professionals broach the topic of modern economic recovery in developing countries. However, this concept, at the time of the writing of this book, has not been a common subject in Liberia. The reason for this may have to do with the fact that more emphasis is placed on telecommunications rather than other facets of ICT. This obvious approach of genuflecting toward the telecommunications aspect of ICT and neglecting the others can, in my opinion, slow, if not stagnate, the growth of ICT as well as economic development efforts in Liberia. Moreover, as the world continues to move toward novel technologies, neglecting any aspect of ICT4D will only result in widening the proverbial digital divide.

This chapter discusses ICT4D and the three categories of ICT. Additionally, the United Nations' ICT4D initiatives through UNDP and the need to establish an agency of government in Liberia dedicated to other technologies will be also be discussed.

To begin, let us define ICT4D. ICT4D, or information and communications technology for development, is an initiative that is used to bridge the digital divide between developed and developing countries and to aid in economic development and to provide up-to-date information communications technologies. ICT4D uses information and communications technologies simply as platforms for economic development. Mainly, it focuses on ICT policies and strategies formulated and implemented for economic development.

At the 3rd IEEE/ACM International Conference on Information and Communications Technologies for Development (ICTD2009) held in Qatar in April 2009, the term "ICT" referred to computing devices (e.g., PCs, PDAs, sensor networks), technologies for voice and data connectivity, the Internet, and related technologies. The domains in which ICTs can be applied, according to the conference, include but are not limited to education, agriculture, enterprise, healthcare, poverty alleviation, general communication, and governance. The application of information technology has been responsible for economic growth in many parts of the world.

To better explain ICT4D three main categories are used: information technologies, telecommunications technologies, and networking/ Internetnetworking technologies.

1. ***Information Technology (IT)***—IT is the use of computers and their accompanying software to manage information. It involves the processing, storage (for retrieval), protection, and transmission of information in a timely and efficient manner. In Liberia, this aspect of ICT has progressed slowly, resulting in a greater marginalization of those without ICT knowledge and a broader national digital divide.

2. ***Telecommunications Technologies***—Telecommunications technologies comprise telephones (land-based, cellular/ mobile, etc), radio, television, fax, and other devices that operate through satellites or other media. The telecommunications sector in Liberia has seen colossal advancement in terms of its penetration into all parts of the country. It has also proven to be very lucrative; hence, the undivided national attention and focus placed on it by the government and the private sector as well as the entire Liberian population. Telecommunications technologies, especially cellular technologies, have had an overwhelming impact on the Liberian society, including its economy. They

have broken several barriers to development, especially geographic boundaries, which prevented the country from exploring opportunities that lead to sustainable economic development.

3. ***Networking/Internetworking Technologies***—The most ubiquitous networking/Internetworking technology is the Internet, which has impacted the world in a revolutionary manner. It leverages Internet protocol to facilitate a plethora of communications technologies that have influenced economic development around the world. Distance learning, electronic commerce, etc., are just a few of the benefits of networking/Internetworking technologies. In Liberia, its impact is being felt as well. The use of electronic mail and social networks is endemic among younger Liberians, while e-commerce, which has the potential to create jobs and reduce the price of commodities, has yet to be explored. But this is due to the fact that the country is still recovering from a prolonged civil war and several other factors. Also, networking/Internetworking technologies are not ubiquitous in Liberia either because most of the technology initiatives are still in their infancy or because some feel that the time is not the right time to explore them. But the question is, if not now, then when?

All of the technologies I just discussed have combined to revolutionize the world and have become crucial to the success of every nation and its people.

The UN (UNDP) and ICT4D

The United Nations, through its development "arm," the United Nations Development Program (UNDP), pioneered and participated in ICT4D for several years. Situated in more than 160 countries, UNDP has been involved in several initiatives that leverage ICT4D in developing countries. The UNDP has been very instrumental in helping countries design strategies using ICT as an enabler and aligning ICT with their poverty reduction strategies (PRS). It has also taken several progressive steps toward the millennium development goals (MDG), which world leaders have pledged to achieve by 2015 (United Nations Development Program n.d.). According to information on its Web site, the UNDP is focused on assisting countries "build and share solutions to the challenges of Democratic Governance, Poverty Reduction, Crisis Prevention and Recovery, Energy and Environment, ICT For Development (ICT4D) and HIV/AIDS" (United Nations Development Program n.d.). UNDP also identified five strategic areas of ICT4D-related interventions as another way of enhancing ICT4D. These five strategic areas are:

- National ICT for development strategies
- Capacity development through strategy implementation
- E-governance to promote citizen participation and government transparency
- Bottom-up ICT4D initiatives to support civil society
- National awareness and stakeholder campaigns

The UNDP created and dedicated an ICT4D trust fund as a flexible mechanism to support the above activities. Apart from ICT4D the UNDP has been an advocate for change and leverages knowledge, experience, and available resources to help better the lives of people (United Nations Development Program n.d.).

There are other organizations that are involved with ICT4D with a commitment that parallels that of the UNDP: the International Monetary Fund, the World Bank, governments, academia, NGOs, corporations, etc., have all engaged in this endeavor.

The Need For An "Arm" Of Government to Focus On ICT

ICT4D uses ICT through set policies and strategies to address several issues in a society: education, business, poverty, innovation, entrepreneurship, development, and so on. To make full use of the benefits of ICT, Liberia must have a balanced approach toward ICT4D. Rather than genuflect toward one aspect of ICT—the telecommunications sector—Liberia must develop and implement strategies that will demonstrate a balanced approach. This can be done by reinforcing the ICT mandate given to existing governing agencies of government.

Another approach could be to establish an agency fully dedicated to ICT. Hopefully, a dedicated ICT entity with particular ICT responsibilities will encourage innovation and create new economic opportunities for Liberia. If this is implemented, Liberia will join the ranks of other developing countries that have taken a similar approach. Nigeria has the Nigeria Information Technology Development Agency (NITDA), while other countries have different names for agencies within their governments dedicated to and responsible for ICTs.

Part IV
Business

Chapter 19

E-Commerce in Liberia: The E-TradeLiberia.com Project

This chapter is to me one of the most important chapters in this book since it provides a complete paradigm shift in the area of commerce and business in Liberia. It discusses an e-commerce scenario using ETradeLiberia.com, a Web portal that I created while researching the possibility of implementing e-commerce in Liberia. The chapter provides a pioneering initiative that utilizes e-commerce to change the way business is done in Liberia. It identifies a better and cheaper medium for Liberians to trade goods and services and offers new ways for small business owners, farmers, and other stakeholders to improve and globalize their businesses. The chapter further identifies the plethora of benefits that an e-commerce initiative can bring to Liberia. In a nutshell, the chapter introduces a framework for e-commerce adoption and advocates a fundamental change in business paradigms in Liberia.

Thomas Friedman wrote in his book *The World Is Flat* that fiber-optic cables and several other technological advancements are responsible for the flattened world in which we currently live. But not only have fiber-optic cables and technological advancements flattened the world, they have changed the way business is done. This has led to what is referred to today as the "new economy." The advent of the so-called new economy driven by the Internet and its technologies provided new windows of opportunities for businesses to transcend geographical boundaries and tap into markets of other countries. E-commerce (EC), a not-so-novel phenomenon and component of the new economy, has also attracted brick-and-mortar businesses as an added value. But despite its successes in developed countries, e-commerce has not been a universal phenomenon. Developing countries, especially those on the continent of Africa, have yet to experience its benefits. Liberia is one of the many developing countries that have yet to adopt e-commerce. This is as a result of a devastating and prolonged (fourteen-year) civil uprising.

The idea of e-commerce in Liberia for economic development remains novel and has only been mentioned briefly in policy documents. The fact that it is crucial to Liberia's global economic initiatives has not been acknowledged by stakeholders, who, it appears, might be discouraged by the challenges involved in its implementation. Hopefully, after learning about ETradeLiberia.com, there will be a change of minds. Before going into ETradeLiberia.com I will first briefly discuss e-commerce.

What is E-Commerce?

E-commerce is the buying and selling of information, products, and services over the Internet (Aida Opoku Mensah 2006). It includes several activities: "trading of goods and services, delivery of content online, electronic fund transfers, electronic share trading, electronic bills of lading, commercial auctions, online sourcing, public procurement, direct consumer marketing and after-sales services" (Aida Opoku Mensah 2006).

E-commerce falls under four main categories: business-to-business (B2B), business-to-government (B2G), business-to-consumer (B2C), and consumer-to-consumer (C2C).

E-commerce has replaced the traditional form of business-to-consumer transaction, which involved the buyer, seller, retail shops, catalogs, etc. Turban et al. (2008) defined e-commerce from several perspectives: From the business perspective, e-commerce is doing business electronically where information is substituted for physical business processes. From the service perspective, e-commerce is used as a tool that delivers goods and services in a cost-effective and fast way. Stakeholders in the market (governments, consumers, firms, etc.) are beneficiaries of this service. From the learning perspective, e-commerce enables online training and education in educational institutions, businesses, and other organizations. From the collaborative perspective, e-commerce provides a medium for interorganizational and intraorganizational collaboration. From the community perspective, e-commerce provides a medium using Web 2.0 technologies for education, transaction, and collaboration. Facebook and MySpace are examples of how e-commerce can be viewed from the community perspective.

Modern firms come in two forms: Those that are purely physical are referred to as brick-and-mortar firms. Those that engage solely in e-commerce are known as virtual or pure-play firms. Firms that use e-commerce as a separate marketing channel are referred to as click-and-mortar firms.

The success of e-commerce is dependent upon access to the Internet. Access to the Internet has been a major topic in discourses involving ICT penetration in developing countries. Since its advent the Internet has experienced significant growth. Less-privileged countries like Liberia, which had in previous years not been able to access the Internet, are now able to do so as a result of improvements in mobile technologies. This improvement paves the way for e-commerce and m-commerce in countries that initially did not have the capacity or readiness to implement an e-commerce infrastructure.

An UNCTAD report of 2001 lists three stages of e-commerce readiness. These three stages are listed below:

- *Readiness*—The first stage of e-commerce development. The readiness of people, business, infrastructure, and the economy is crucial at this stage.

- *Intensity*—The second stage of e-commerce development deals with the intensity with which ICT is utilized within a country. It also involves the extent to which e-commerce activities are undertaken.

- *Impact* —The last stage of e-commerce readiness is the degree of impact on the national economy and business activities.

The above three stages listed by UNCTAD are crucial to an e-commerce implementation.

E-Commerce In Liberia

In Liberia none of the stages mentioned above has been achieved. The relative progress made in ICT has not included e-commerce. Business continues to be done in the traditional way, which denies the country of potential opportunities that could be gained from the global market. Although there are a few Liberian-owned Web sites, such as TLCAfrica.com, CallLiberia.com, LiberiaObserver.com, FrontPageAfrica.com, RunningAfrica.com and others, that provide e-news, advertisements, and e-entertainment service online. None of these firms operates as a pure-play e-commerce Web site that provides secure online transactions with financial institutions in Liberia.

Awareness of the impact of e-commerce is still lacking on several parts of the African continent. Moreover, the opportunities offered by e-commerce are yet to be acknowledged by many African countries (Aida Opoku Mensah 2006)

Liberia obviously is one of the African nations that lacks awareness of the impact of e-commerce.

E-TradeLiberia.com

ETradeLiberia.com is a novel and strategic approach to economic and social development in Liberia. At a time when digital technology has transcended the geographical boundaries between countries and significantly changed the global supply chain, ETradeLiberia.com provides an option that enhances Liberia's entry into the "new global economy."

A discourse involving e-commerce in many parts of Africa might be met with cynical and pessimistic responses. But for a country undergoing recovery, bold and pioneering initiatives need to be explored in order to find better solutions. ETradeLiberia.com presents what appears to be a pioneering initiative for Liberia. It involves the use of e-commerce in Liberia to improve commerce and business transactions as well as an attempt to globalize the country's economy. ETradeLiberia.com illustrates that there is a need for e-commerce in Liberia to support the agro-business sector and small and medium business as well as solve the problem of real estate (crowded markets) that small businesses and struggling entrepreneurs face.

Figure 19.1. A snapshot of ETradeLiberia.Com

Explanation Of ETradeLiberia.com

As mentioned above, the overall goal of ETradeLiberia.Com is to present a case for the adoption of e-commerce in Liberia as a part of an economic development initiative. The project was Web based and included online transaction arrangements with banking and financial institutions and the incorporation of m-commerce technologies. The participants are expected to be stakeholders of Liberia. The Internet and its protocols along with other Web applications were leveraged, and the domain name "ETradeLiberia.com" was selected to illustrate its virtual nature. Outcomes from this initiative are listed as follows:

1. Establishment of an Internet-based medium that exposes Liberian markets/businesses to the global economy

2. Level the playing field for farmers, market women, small/medium-scale entrepreneurs, and exporters to be able to trade their products locally and globally

3. Drastically reduce the effect of poverty in Liberia by providing a medium for small-business owners who do not have the real estate to run a brick and mortar business

Payment Method—Ecobank's Visa Electron Card

The use of credit cards and other forms of payments for online transactions has reached Liberia, although not quite ubiquitously as it has in other countries. In fact, it is still in its infancy. Online payment is one of the imperatives of an e-commerce initiative, and this has proven to be a major challenge, not only in Liberia but in other African countries endeavoring to adopt e-commerce. The introduction of the Visa card and its technologies in Africa provides hope for an online payment system. As I mentioned, Visa cards have reached Liberia and are being used in certain areas. The Robert L. Johnson Kendeja Resort in Liberia provides options for the use of credit or debit cards.

Ecobank, one of Africa's premier banks, offers several banking options in the West African subregion. One of those options is the use of Visa Electron cards, which is the selected option used in this initiative. The Visa Electron card allows customers in eight West African countries to access more than 16,000,000 point-of-sale terminals and over 840,000 ATMs around the world (Ecobank n.d.). The Ecobank Visa Electron and debit card can be used for payment of goods and services at commercial facilities carrying the Visa Electron signs visibly. The card can also be used on a twenty-four-hour basis. The problem with the Visa Electron card is that its availability is limited in Liberia, and this presents more challenges to an e-commerce initiative. But progress is being made in Liberia's financial sector that will ultimately allow a better form of online payment to be used in the country's e-commerce initiatives.

Mobile technologies, which have higher penetration in developing countries, continue to enhance m-commerce. For example, in Liberia, the International Bank in 2008 introduced and launched the small message services (SMS) for its customers. This feature allows customers to access their bank account using mobile phones or m-commerce (Ecommerce Journal 2008). Other countries including South Africa have new and unique methods of online/mobile payments. For example, WIZZIT, a banking firm in that country (South Africa), provides an online banking transaction mechanism that is done via mobile phones with the ability facilitate e-commerce (Katz 2006). Zambia National Commercial Bank (ZANACO) and other banks in that country have been modernizing their services by offering their customers e-commerce services (Malakata 2005).

E-COMMERCE WORKFLOW—CUSTOMER TRANSACTION

Here is how a regular transaction session is done on ETradeLiberia.com. The process begins with the seller advertising the item to be sold on the ETradeLiberia.com Web site. The customer visits the site, locates an item, and decides to purchase it. The customer then places the order using the site's shopping cart. The order is processed via the bank (Ecobank or any other financial institution). A process-completion notification will be sent to the buyer and seller as soon as payment processing is done. Information from this process is also stored in an E-TradeLiberia.com customer-information database. The notification process "triggers" the seller/storage to ship or deliver the item to the customer. The transaction ends when the customer receives the item.

THE M-COMMERCE OPTION

The m-commerce option follows the same workflow pattern discussed above, except mobile devices are used in place of computers. Mobile phones are used to perform business transactions over the Internet. M-commerce has garnered tremendous attention and will continue to be the focus of ecommerce adoption in developing countries.

Both e-commerce and m-commerce may apply differently in Liberia's agricultural sector. There may be a slight difference in the delivery process as not many farmers in the rural areas have the capacity to make deliveries. In this case, the customer will pick up the item instead of the seller performing delivery.

Challenges And Benefits Of E-commerce In Liberia

The slow pace of ICT penetration, the lack of infrastructure, and several other factors have hindered e-commerce adoption in Liberia. Moreover, the Internet-savvy portion of the population consists primarily of young unemployed men and women who cannot afford to make purchases or own a Visa card needed for an e-commerce transaction. Payment methods in most parts of the country are still largely being done on a cash basis. Only an optimistic view of the country's economic and political future would take into consideration these young people as potential customers. Additionally, the roads that connect various parts of the country are dilapidated and will present a serious problem for the delivery of goods to the rural areas. Other challenges that need to be addressed before e-commerce can become fully operational are taxation, security, usability in an illiterate society, sustained political stability, and Internet access.

The implementation of e-commerce in Liberia, I believe, can provide more than five hundred to a thousand new jobs and most and of all can improve the economy. When it succeeds, Liberia will be a changed country. Streets will not be as congested as they currently are because vendors will use cyberspace to market their goods. The quality of goods will be better because of competition, and farmers will have a medium to expose their products to a larger market. With unemployment at astronomical heights, Liberians have been forced to engage in small businesses or small, medium, and micro-sized enterprises (SMEs). These businesses have the potential to drive the Liberian postwar economy if they (businesses) are approached in ways that are unlike the traditional ways on which Liberia is so dependent.

Building on the microfinance initiatives such as those used in Cameroon, Liberia, can leverage e-commerce/m-commerce to reduce poverty and enhance the living conditions of its people (*Red Herring Magazine* 2001).

The E-Commerce Initiatives Of Liberia's Neighbors

Several countries in Africa, especially West Africa, have already engaged in e-commerce. Ghana recently initiated online commerce using eTranzact, which allows the transfer of money through mobile phones and the Internet (ETranzact 2009). There are several other e-commerce activities taking place in Ghana as well. In Nigeria, e-commerce adoption is slow but steady. Online money transaction has been the main form of e-commerce activity occurring in that country (Economist Intelligence Unit 2006). Sierra Leone, which closely borders Liberia, has also engaged in some form of e-commerce activity through the SMS-based mobile payment service that was launched by MoreMagic and Splash (E-commerce Journal n.d.). There are several other African countries that are in the process of adopting e-commerce, although not as aggressively as Western countries.

CHAPTER 20

BUSINESS EVOLUTION: THE GENESIS OF THE "DIGITAL ECONOMY" IN LIBERIA

In a digital Liberia, businesses that align their practices with the new paradigms of economics, by leveraging ICT, will run more efficiently and have competitive advantage in their respective markets. Any business that fails to adapt to the trends of the digital economy runs the risk of going out of business. Times have changed, and economics is changing (although the fundamental principles remain); so too should businesses. This chapter talks about the evolution that has begun to take place in the business sector of Liberia. It focuses on the ATM machine, which leverages electronic financial transactions, and its penetration in the country as well as the business opportunities it will provide for the Liberian people. I discuss new entrants into the business sector, specifically small businesses and how they can compete with bigger businesses due to the benefits the Internet provides. I also discuss the need for businesses to reengineer their strategies in order to be able to compete in the digital community. Finally, I address some of the challenges of this evolution and provide some solutions.

Liberians will embrace pioneering initiatives as long as those initiatives bring development and opportunities to the country. Hence, it was no surprise when I learned that Liberians had overwhelmingly embraced Ecobank's introduction of the ATM card (Cephas 2009). Ecobank's initiative demonstrates that to succeed in the "new economy," firms will need to have vision, be innovative, have the audacity to explore the unfamiliar, and pioneer new and exciting ways to do business in order to gain durable competitive advantage. Success in this global and digital economy requires a departure from the status quo, which means either reengineering processes or exploring new and innovative approaches to business.

The Automated Teller Machine In Liberia

The introduction of the ATM card kindled the genesis of a digital economy in Liberia. Pioneering and innovative business initiatives will be established because of the ATM card. And I believe that Liberians will take advantage of the opportunities that the ATM card brings. This will give them control of the economy and enable them to compete globally.

To me, an ATM, as it is widely referred to, is just a computerized money-dispensing machine that is connected to banks through a telecommunications medium. Or in layman's terms, an ATM is a digital bank teller! But to give a more textual definition, an automated teller machine is a terminal used by most banks to provide their customers access to financial transactions in a public space without the need for human representation or a bank teller. Customers are issued a plastic card on which their banking information is embedded to be used as a form of identification when using the ATM machine. Hence, when a customer accesses a terminal, which is often placed in a public area, the ATM calls the bank's computer system to verify the authenticity of the customer as well as the available balance.

The ATM is a computer with additional technologies that allow it to read cards, process transactions, dispense money, print receipts, and perform numerous functions. It has a central processing unit (CPU), a display, a printer, a secure cryptoprocessor, a money dispenser, a card reader, etc. ATM machines also use operating systems such as Microsoft Windows and Linux, which makes them comparable to the computers we use in our offices and homes. They use broadband (some use dial-up) connections from the point of location to communicate with the banks' computer systems.

ATM machines are not only used for dispensing money, though. They can be used to purchase tickets, check bank balances, and make money transfers; at universities or other institutions of learning they are used to output transcripts for students, just to name a few uses. The multifunctional capabilities of the ATM machine makes it ideal for businesses, and this is why I believe it will pave the way to a plethora of business opportunities for Liberians as we move forward.

ATMs are situated in places other than banks, places where they can be accessed by a large number of people. In Liberia, it would be prudent for businesses and financial institutions to situate ATM machines in places like Roberts International Airport, Broad Street (under the Ministry of Education), Waterside, the Freeport, the Port of Buchanan, in government ministries, at universities, grocery stores, convenience stores/gas stations, night clubs, restaurants, etc. ATM machines are usually connected to a network or a conglomerate of banks. This allows customers or anyone with an ATM card to fully utilize that card at any ATM machine as long as he or she has an account with an existing bank that is connected to the "interbank networks." This is one reason why banks in Liberia will need to work together in order to reap the benefits of this relatively new approach.

The use of an ATM card often requires the payment of a fee, depending on the bank. Some banks charge certain fees, while others don't. I remember that during my trip to Liberia in December 2009 I used my ATM card at the Ecobank's terminal on Randall Street. I did not realize the fees charged until I returned to the United States. Although they were not exorbitant, I don't remember the ATM machine in Liberia cautioning me about the fees that would apply as is done in the United States.

In the case where an ATM is owned by a nonbanking institution or an individual, a service fee will most likely be charged. Most often, the money from the ATM is not drawn from a bank but rather the vault of the owner or leaser of the ATM machine. For example, let's say I own a convenience store in Caldwell that houses an ATM machine. When a customer comes in and needs "fast cash" he/she can use the ATM machine in my store to make a withdrawal, obviously for a fee. Alternatively, if I have a "point of sale" system set up, the customer can swipe his/her card at one of my terminals and select the amount of cash back needed. Again, this is done for a fee. Simply put, Liberians can also get into the ATM business to a make profit. I discuss that in the next section.

A Business Model That Includes The ATM Card

ATM Network, Inc. is an ATM company that provides full-service ATM programs, including buying, leasing, or renting out space to an ATM. A while back, I made several calls to this company to garner information about their ATM machines. Based on the information received, I surmised that the advent of the ATM in Liberia indeed presents several unprecedented opportunities for Liberian entrepreneurs who are willing to exert their creative energies to pioneer new business endeavors. ATMs, as I said previously, can be owned by anyone interested in leasing, renting, or situating them at their location for customers' use. As Liberia moves forward, Liberians will begin to take advantage of the opportunities that the ATM card brings in variety of ways. Grocery stores, gas stations, restaurants, Samuel K. Doe stadium, the Sports Commission on Broad Street, Roberts International Airport, the Freeport of Monrovia, and all those locations that cater to a large consumer base will have ATMs situated there. Universities in Liberia will also have ATMs situated on the campuses for the convenience of faculty, staff, and students. Currently, there is an Ecobank branch situated on the main campus of the University of Liberia. During my visit there, I did not see an ATM machine. But as time goes by, I am optimistic that banks will deploy ATM machines there soon. Most likely when the Fendell campus is completed it will have a plethora of them as well. Ministries and other institutions of government will have ATMs situated at their facilities, and banks will work with the government, the public, and private sectors and NGOs to develop some sort of direct-deposit mechanism that will allow workers to utilize ATMs instead of standing in long lines and risking their safety/security.

To gain competitive advantage in this "new economy," grocery stores, supermarkets, restaurants, and all those who provide goods to their consumers will have to implement a POS (point of sale) system with card-processing functions where consumers will be able to use their

ATM cards if an ATM is not available. Point-of-sale systems are used in retail operations to process transactions upon the consummation of a sale using cash, checks, credit/debit cards, or any "legal tender" that is accepted by the seller or service provider. Information from that sale goes into computer systems/databases. Think of it as an electronic cash register that allows you to use an ATM card.

Not to be redundant, but all of these will work better if broadband connectivity gets to Liberia. At present, this can work with the current infrastructure, but it will be more efficient and widespread with low-cost and high-quality Internet connectivity. Point-of-sale systems, like the ATM system, require connectivity to banks. Obviously, some areas have begun to apply these types of business transaction approach, which is a result of the progress being made by LIBTELCO, the plethora of Internet services providers, and the countless VSATs (very small aperture terminals) providing connectivity to LANs/WANs and the Internet. But again, to be ubiquitous and efficient, broadband will be needed.

Liberians now need to use their creative energies to enter an ever-changing digital economy and must be ready to face its challenges. This is where Liberian ICT professionals can give back and move this country toward a new economy. The best method for tackling the challenges of this new economy and finding solutions is to take a positive, optimistic, innovative, and collaborative approach.

The Internet Has Created A Level Playing Field

Da Costa (2001) notes that small companies are increasingly becoming powerful because of the Internet and e-commerce. He further notes that the phenomenon that allows small businesses to become powerful has also made them creators of jobs and drivers of innovation and economic development all over the world.

The ATM will also change the way we transact business in Liberia. As we move forward, businesses will be transacted over the Internet using the ATM card through online transaction processing (OLTP) mechanisms. With this as a payment method, brick-and-mortar stores will not be as necessary for entrepreneurs. Virtual stores will be opened on the Internet, thereby eliminating the need for a physical sales location; but this will increase the need for storage. This is because sellers will store their products at a home or in warehouses and list their products on their Web sites. Buyers will access the Web site, select the products of their choice, and either pay online (if they have confidence in the system) or call the seller and have product delivered for payment upon receipt. This will eliminate the cost of real estate (stores) for the seller and alleviate the problem of street vendors (if it's not already prohibited), which has become a major problem for city corporations in Liberia. It will also be convenient for the buyer because he/she will not have to commute to a brick-and-mortar store to purchase goods when the transaction can be done by the click of a mouse and delivered to him/her. Large companies could create courier services to facilitate this initiative and through that create more jobs.

When the appropriate mechanisms are implemented and leveraged, and if banks in Liberia align their processes with global institutions (Visa, MasterCard, etc), e-commerce and online banking will also greatly benefit Liberians living in the Diaspora. Liberians in the Diaspora who are struggling to build new homes in Liberia will be able to use (responsibly) credit (credit card) opportunities to purchase goods such as building materials over the Internet. Stores in Liberia will have to create e-commerce Web sites to compete in this environment. Now, while this alleviates some of the problems we (those living in the Diaspora) face with using proxies to implement projects in Liberia, it is certainly not the ultimate panacea. Physical presence, in my opinion, is the solution even though it requires spending around $2,500 USD for a round-trip ticket and spending weeks away from work.

Banks will have to adapt to the new paradigms of financial transactions, which are now defined by the flow of electrons and information. Currently (2010), Ecobank in Liberia is in my opinion the leading bank because of its willingness to be innovative and its courage to explore new, exciting, and efficient ways to do business. As mentioned above, firms or businesses that are slow in adapting to the new paradigms of business will witness a mass exodus of their customers to companies and banks like Ecobank. This relatively "new" approach to banking has opened the way to an irreversible culture that will be a part of Liberia's transformation—from a country held captive by poverty, illiteracy, and high unemployment rates to a country that is economically viable.

Business strategists and ICT professionals who plan to be involved in this process must be able to deliver solutions. Today's work environment demands vision, creativity, and intelligence in order to attain market advantage. I often remind my colleagues that our jobs have a great impact on our firms' bottom line; that we are defined by the way we align technology with the strategies of the firm. We are not defined by how quickly we fix systems or how "techie or geeky" we look. Aligning ICT with business strategies to bring added value and competitive advantage are the things our employers look for. They (employers) want solutions, not problems, and they want to know what we are capable of doing. That's our job; to provide solutions ... and at low costs.

There will be a great demand for business strategists and ICT professionals to implement these new approaches, especially for business-to-business (B2B), business-to-consumers (B2C), and business-to-government (B2G) transactions. Web developers will be needed to design Web sites, and those with a background in e-commerce will be aggressively sought. Business students (economics, accounting, management, etc.) will have to learn, understand, and adapt to the dynamics of the "new economy." But the most significant task is educating the Liberian people about the benefits these new changes will bring.

Liberia will be a success story. In fact, just as other countries rose from war to achieve sustainable economic growth, so too will Liberia rise from its current situation. But to achieve this, Liberia will need leaders with vision, creativity, and passion. The burden of Liberia's recovery and economic growth rests on those of us who are in the Diaspora. Liberians cannot continue to rely on our friends (donors, philanthropists, humanitarians, etc.) for help all the time. We have to do it ourselves! This is obviously the reason why those of us in the Diaspora cannot make obtaining a job in the government a condition for returning home to contribute to the recovery of the country. Getting a job to sustain one's family while positively contributing to the recovery process is understandable; requesting funding from government, an NGO, or the private sector to fund a project is also understandable. But requesting/demanding a position in the government, as some of my contemporaries are doing, should never be a condition for returning home to rebuild one's country. As educated men and women we should know that getting a government job is not a right but a privilege. My parents paid their share of taxes, and I am an educated man, but that does not automatically guarantee me a job in government or with a public or private firm. Contribution to the successful development of Liberia *must* be unconditional and unwavering.

Despite the benefits that are involved in an "evolved" business environment, there are other challenges that will need to be addressed, such as the issue of security. The government of Liberia's security apparatus will need to be trained in cybersecurity and be more alert in terms of providing physical security. Financial institutions that own ATM machines situated in open locations must install alarm systems and apply needed security systems. Security administrators and all of those IT professionals who administer information systems or were hired to provide security must be smart, vigilant, and conscientious in providing security to their employers as well as their employees. They will need to educate their stakeholders and remain trained and up-to-date on measures needed to counter any threats, be they cyber or physical.

As I mentioned above, those entrepreneurs who would like to expand their business operations and ventures will have to adapt to the dynamics of the digital economy, and this requires establishing an online presence. I strongly believe that banks, insurance companies, pharmacies, and companies like Wazni Bros, Fouani Bros, Petrol-1 Inc., and several companies and organizations operating in Liberia will gravitate toward an online option to buttress their standing among competitors. But my advice to them is to exercise some level of caution when selecting a service provider in this ultra-competitive market, especially if they are not running their own platform. Many companies and individuals will take advantage of innovative technologies (cloud computing) to exist as service providers. They will want to enter key market segments that are good for competition. The problem with this is that not all of them may have the experience, resources, and technical expertise to offer a value-added service. And in a competitive business environment, the last thing a firm or business organization should want is a diluted or ineffective online presence.

Capacity will grow as Liberia integrates technology into its educational system and Liberians in the Diaspora return home to take advantage of this opportunity. But as the digital market matures in Liberia, additional skills and expertise will also be needed, especially from expatriates. Every Liberian knows that most developmental initiatives do come about from "partners in progress," usually of foreign backgrounds.

Political will is crucial! The success of the business evolution in Liberia depends on support from those in power. The government of Liberia as well as other stakeholders will have to provide the support for this new business paradigm in order for it to be vibrant and successful. Once this shift in business paradigm occurs, businesses in Liberia will forever be changed. For this reason, everyone must be ready to adapt and be prepared to face the potential challenges of a digital Liberia.

Part V
National Security

CHAPTER 21

CENTINOL—A MODERN NATIONAL SECURITY INFORMATION SYSTEM THAT LIBERIA SHOULD ADOPT

This chapter is based on a research project that I implemented during one of my graduate programs. Parts of it were also published on the *Liberian Daily Observer*'s Web site sometime in September of 2009. The publication of that article generated a lot of questions and suggestions from the *Daily Observer*'s readers, who feel that national security has been and is still a major issue. The responses I received in reference to the article indicated that Liberians are willing to be involved in the "re-creation" and modernization of their national security sector at all costs. From what I surmised, though, the desire for this endeavor stems from their experiences during the Liberian civil war and their unwillingness to have the same repeated.

Sometime in 2009, the Liberian *New Democrat* newspaper reported by Liberian security forces and heralded in local print and electronic dailies that some Pakistani nationals, suspected of being terrorists, were arrested at the Roberts International Airport after being caught with fake passports (Roberts 2009). This news raised some concerns in the Diaspora since prior to that incident, the U.S. government's Department of Homeland Security had reported that terrorists might be seeking refuge in underdeveloped countries to plan and execute their operations. The apprehension of the suspects and subsequent incarceration by Liberian security officials is not only indicative of the changes that have occurred in Liberia but also a renewed and rekindled exhibition of vigilance and conscientiousness. Previously, the suspects might have evaded arrest through bribery and other devious means. What makes this uncommon and interesting is that in the past, Liberian security officials like those in other underdeveloped countries, have been reported to be susceptible to bribery and other corrupt practices.

Yet despite issues faced by the country's security sector, some changes have been made and are worth commendation. There is now a newly trained national police force and a new national army, and other arcas of the country's national security apparatus have also experienced some changes since the cessation of hostilities.

But the "Pakistani" incident and other incidents that occur daily in Liberia illustrate the need for continuous reforms in the Liberian security system. The fact that the suspects made it onto Liberian soil indicates the need to implement mechanisms that provide our national security officials the intelligence needed to combat twenty-first-century crime. Nowadays, it is obvious that criminals are juxtaposing modern approaches with their traditional modus operandi to unleash crimes on societies.

To enhance the country's national security endeavors, I proposed CENTINOL (Central Intelligence Network of Liberia), which is a national security information system. CENTINOL was originally conceived by me but was later updated in collaboration with a few ICT professionals whose names and area of contribution/expertise I have listed below:

- Marvin Cassell—Telecommunications
- Kamara Watson Jr.—Information Security
- Kassa Cooper—Networking/Data Center Administration
- Emmanuel Toe—Database Administration
- Charles Johnson—Hardware support

I remained the owner and contributed toward the programming and Web development aspect of this research endeavor. Before proceeding, let me clarify that I am not an expert in national security, nor do I aspire to be. I am, however, an ICT professional with a background in information security and am familiar with the impact of ICT on national security. CENTINOL was conceived and researched from an ICT perspective. Overall the consideration and implementation of CENTINOL will require the involvement of several stakeholders, including national security experts.

Liberia's Poverty Reduction Strategy (PRS) paper identifies several crucial areas to be addressed and delineated actions to be taken to bring development to the country. National security was one of the areas identified in that document as one of the top five critical sectors requiring immediate recovery. Several strategies and plans were devised to restore the national security sector, but none of them involved the implementation of a national security information system that enhances collaboration and information sharing. Currently, the country's national security sector is being supported by the UNMIL, which has a mandate to be in the country for a few more years. Since the Liberian government is expected to take over the security of the country upon the departure of UNMIL, the need for a modern national information security system that can be run by Liberians has become imperative.

This chapter addresses that need by proposing Central Intelligence Network of Liberia, CENTINOL. The acronym CENTINOL is pronounced in a way similar to the word SENTINEL, which means sentry or guard. CENTINOL is being proposed as a framework for a modern national security information system that will enable collaboration and information sharing among security agencies (police, immigration, NBI, national defense, Customs, SSS) in Liberia, with link to the INTERPOL database. INTERPOL, the acronym for International Criminal Police Organization, is an international intelligence agency permitting collaboration among intelligence agencies around the world. CENTINOL will host a gigantic and secure database or data repository that links to every government security agency in Liberia, all Liberian embassies granting passports and visas to Liberians and foreign nationals, mobile security personnel stationed at the various ports of entry, and INTERPOL.

The development of CENTINOL as a secure system to be linked to a secure database is critical and must be primary. Secure systems development and paradigms are discussed in this chapter to cover that aspect. CENTINOL software, which will provide real-time and accurate intelligence to security officials through searches and real-time notification, is discussed as well. There is also a brief discussion of the mobile client that will allow wireless communication technologies. Wireless access will be incorporated in CENTINOL for mobility. A Web-based option is discussed as well as a mechanism that will allow interoperability among disparate systems. Ultimately, CENTINOL aims to be "an information system that allows the use of a variety of devices and systems, including mobile phones, tablet PCs, personal digital assistants (PDAs), and laptops equipped with wireless LAN (Local Area Network) or General Packet Radio Service (GPRS) to access data in the central database" (Wilkins, *Liberian Daily Observer* 2009d).

Also covered in this chapter are the principles of secure software programming and the future addition of needed functionalities to CENTINOL.

Central Intelligence Network Of Liberia

National security is critical to every country and must always be fully supported by a country's government. In the twenty-first century, national security initiatives must adopt new paradigms, especially those that subsume modern criminal activities and modus operandi. This means that information systems must be developed with security as a priority and based on the security development lifecycle and secure software-development practices.

The major functionalities of the CENTINOL system are collaboration and information sharing. Components and mechanisms that will enable collaboration are e-mail, phone, video, fax, notification, and so on. The goal is to provide real-time information from CENTINOL's database to law enforcement officers to efficiently carry out national security operations. The system will be able to differentiate between the levels of information sharing among law enforcement officers. CENTINOL will provide real-time alerts via portable communications devices such as PDAs and cell phones. It will also provide a mechanism for privacy protection. A Web-based component enables CENTINOL to provide universal access to users and law enforcement officers. CENTINOL borrows from the approach used by Chen et al. (2002) by using three-tier Web application architecture. The choices of programming platform used include Java, JSP, and HTML. A JDBC connection to the database will be used for access to backend data. The business logic will be coded in JavaBeans for configuration and extensibility as proposed (Chen et al. 2002).

In CENTINOL, security officials or users will have to be authenticated or verified before they can access any data. Authentication occurs when the user provides credentials such as user name and password, fingerprint, smartcard, etc., at login or at a security pad located at the point of entry. If or when the user is authenticated, he/she is granted access based on rank or security clearance. Permission to files will vary as only certain users will be able to access, add, edit, or delete data; other users will have read-only access. This process allows for internal checks and balances by monitoring the movement and actions of all personnel using the system, making it easier to determine any breach from within and prevent those from outside.

The CENTINOL System

There are two main components of the CENTINOL system: the CENTINOL database and the CENTINOL software.

The Database

The CENTINOL database is a robust repository of information that is efficiently organized in a manner that facilitates access, ensures flexibility, allows faster retrieval of information, and provides an expanded search capability. It serves as the central records management system (RMS) that stores data input from the CENTINOL client entered by the user. The data inputs are culled from criminal case information, information gathered from embassies around the world from those applying for passports, visas, and other events. More importantly, resources provided by the Internet are also integrated to allow better sharing of information and collaboration among law enforcement officials despite geographical boundaries and diversity of the joint security consortium. It also provides access to data to all security agencies in Liberia. It connects to the front end of the system (CENTINOL software) through a Java Database Connectivity (JDBC). MySQL database server, which is an open source database server available for a wide variety of platforms, can be used for the database. The JDBC API is selected because it facilitates the writing of platform-independent Java programs that can be used to manipulate data in SQL databases without any further modifications.

CENTINOL Software

The CENTINOL client software consists of three modules: the desktop module, the Web-based module, and the mobile module. The desktop module is installed on computers that are stationed in-house; the Web-based module provides universal access to the system via a Web browser, and the mobile module also provides universal access and mobility via mobile devices to law enforcement officers who are assigned in remote areas where there is limited IT access. CENTINOL software is a three-tiered application architecture initially proposed by Chen et al. (2002) and will allow users access through a user name and password-based authentication mechanism (Chen et al. 2002).

The Java, JSP, HTML, and XML programming languages have been proposed to be used because of interoperability requirements since CENTINOL connects to disparate systems.

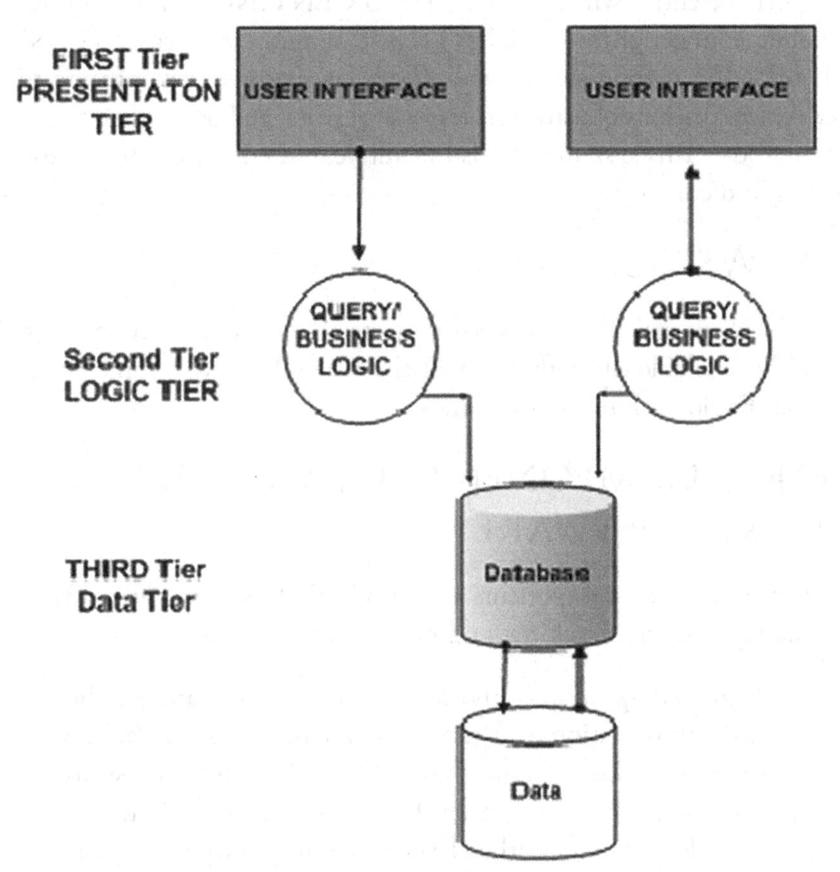

Figure 21.1. A Three-tier Architecture

Business Logic

CENTINOL's business logic "or the functional algorithms that handle information exchange between a database and a user interface can be coded in JavaBeans or through one of the many available options" (Chen et al. 2002). But I propose JavaBeans because it presents opportunities to allow system configurability and extensibility.

Mobile Client

Law enforcement officers located in remote areas, especially at borders and ports of entry, will require access to CENTINOL. This requires a mobile connection to the CENTINOL system via a secure (VPN) network. The mobile system will allow law enforcement officers to receive notifications of warrants, terrorist threats, and other operational information. This system will also facilitate and enhance information sharing and collaboration.

The Web Client

The Web client will allow universal access to all law enforcement officers. It will be available through many of the available browsers, including Internet Explorer, Firefox, and Safari.

Major Components Of CENTINOL Client Software

There are four key components of the CENTINOL software. They include login, query, tracking, and notification modules.

- ***Login component***—The login component manages the login information and uses a user name/password/access level authentication mechanism to allow user access to the system. The "access level" requirement is included to take the user to another level depending on permissions, job title, and description and for identity challenge. The identity challenge is a certain "code word" that is given to law enforcement officers for authentication prior to entering a classified area of the system.

Central Intelligence Network of Liberia

Name:

PIN:

Dept::

Access Level Class A

Central Intelligence Network of Liberia

Designed by: Darren Wilkins

Figure 21.2. CENTINOL's login interface

- ***Query component***—The query component allows the user to perform searches based on level of permission (to protect privacy). Queries produce results on passport number, date of birth, photo, last ten years' record, etc.

- ***Tracking component***—This module is specifically included to allow law enforcement to track individuals traveling to and from Liberia as well as view a "wanted" list of international criminals posted by INTERPOL.

- ***Notification component***—The notification module includes a mechanism for providing real-time information to law enforcement officers based on alerts, both national and international. The alert includes information of the "suspect" with a link to the CENTINOL database with detailed information.

CENTINOL connects to the INTERPOL database through JDBC, allowing data flow between the two entities. This allows law enforcement officers in Liberia to make queries and receive responses from both CENTINOL's and Interpol's databases.

Design Approach

Developing a secure software system requires a diverse group of IT and security professionals. The development of CENTINOL requires a team that includes project managers, national security officials, developers, IT security professionals, and all stakeholders involved in the development of secure information systems. Developing secure systems requires the use of modern and robust penetration test tools and risk-management procedures. The government of Liberia, which will be responsible for the development of the system, will need to establish an environment that is dedicated to the development of secure software systems. The use of external testers must be encouraged to ensure that the CENTINOL system is developed and tailored to suit national security needs (McGraw 2009).

Finally, since CENTINOL is to be used for national security operations, then it is imperative that the entire system (CENTINOL), including the software that runs it, be secure. Software design involves three main groups: the user, the application developer, and the API (application programming interface) developer. In the past, developers created software from scratch. Today, there is reusable software and programming tools such as libraries, modules, and existing components that can be combined to make applications.

CENTINOL's development will require an environment that advocates software design with security from scratch if not from a highly secure pre-existing solution. Several security-focused activities must be included in the development of the CENTINOL system because of its nature: collaboration, diversity, dynamic tasks, and collaboration with the international community. Future enhancements to CENTINOL should include the incorporation of biometrics and body area networks. Obviously, that would take some time and work to do; but it can be done.

CHAPTER 22

CYBERSECURITY: INTERNET CAFÉS AND THE POTENTIAL THREATS TO NATIONAL SECURITY

Internet cafés in Liberia have helped to bridge the digital divide between the country and the global community. Since the implementation of satellite communications technologies, several individuals and businesses have purchased and installed VSATs (very small aperture terminal) to run Internet cafes and other business enterprises. Internet cafés are locations that house several computers that are made available commercially to the public. While they contribute toward the economy and greatly enhance ICT penetration, if not properly secured, they pose a potential threat to national security by virtue of the easy access to the computer networks and the Internet they provide to the public.

This chapter discusses cybersecurity and its impact on Liberia. Focus is placed on Internet cafés and the impact they have on the national security of the country. The supine approach to computer and network security taken by those who are responsible for maintaining networks, Internet cafés, and other areas that provide Internet access to the public is discussed. I conclude by providing security measures that can be taken to prevent and mitigate, if not eradicate, these threats.

During the year 2009 Liberians witnessed several events that demonstrated the country's vulnerability to cyber attacks. One of those events occurred at the Central Bank of Liberia which involved the forging of signatures in an attempt to withdraw $1 million. That incident was a wakeup call to all Liberians to be more conscientious about the existing and emerging security threats to the Liberian economy. Fortunately, the attempt was thwarted and the culprits were arrested. The culprits in the "million-dollar scam" used an old modus operandi—forgery—and they were almost successful. But we are now in the information age, which has changed our cultures. Nowadays, robbers do not need physical access to banks, they can have virtual access to any financial institution that has a computer system connected to the Internet. A twelve-year-old child can rob a bank from his bedroom, his parents' garage, or even from an Internet café located adjacent to the bank with just a few computer clicks or commands. Indeed, times have changed, people have changed, things have changed and cultures have changed.

At a time when more advanced tools such as the Internet and other technologies are freely available and considering the avalanche of cybersecurity issues affecting financial institutions around the world, the need to make some adjustments in our security mechanisms cannot be overemphasized. These adjustments will allow security personnel to adapt to all potential changes in the modus operandi of criminals. Furthermore, the proliferation of Internet cafés in Liberia and the lackadaisical approach to cyber threats taken by those responsible for them make computer systems and networks in the country vulnerable to attacks. The passive attitude shown by attendants of Internet cafés makes it easy for cyber criminals to use their facilities to launch attacks such as malware injection or denial of service (DoS) on national institutions. Denial of Services or DoS is a situation in which cyber criminals make computer resources unavailable for computer users. There is a possibility that the owners of these cafés may not be fully aware of such threats. They may be merely following a business model and not realize the threat their entities pose. But be that as it may, it is still their responsibility to be aware of potential threats and attacks that may arise from of their establishments.

Internet cafés and other areas that provide public Internet access have significantly impacted Liberia's recovery process. They have provided new business and employment opportunities for Liberians and more importantly, they have enhanced ICT penetration which helps close the digital divide. But despite these benefits, they can also be the conduit through which the transmission of malware, fraudulent acts, and destruction of critical information systems can be perpetrated. The threats that I previously mentioned can lead to economic or security problems for any entity, including the Liberia government. However, these attacks can be prevented by instituting some fundamental security measures in areas that provide public Internet access as well as other computer networks in Liberia.

My focus on Internet cafés in this chapter is directly linked to the experiences I garnered during my several visits to Liberia. In January of 2008, while I was in Liberia working on the B. W. Harris School's Smart Technology Project, I visited an Internet café and immediately noticed the relaxed security atmosphere that existed. An attendant who spent more time collecting money than monitoring his café stood by while the following occurred: users were plugging in USB flash drives and surfing various Web sites that were obviously "malware distributors," keyloggers, etc. Worst yet, the attendant charged Internet users who had their personal laptops a fee to connect their laptops to his network. After seeing all of this and having some experience in computer network and information security, I decided to investigate whether there were any security measures taken to prevent the spread of malware or network intrusion. Since the operating system was Microsoft Windows XP, my desire to investigate was immediately kindled. This is because hackers and malware writers take more pleasure in attacking Microsoft's products as compared to those running on other platforms like Linux or Mac.

My security check was basic: I opened the Windows Task Manager to see what programs were running in the background; as expected, I discovered a host of malware programs. While there are several malware programs on the Internet, two of them have been proven to be very common and can easily be propagated: computer viruses and spyware. A computer virus is a computer program that basically spreads across computers and networks by surreptitiously making copies of itself. Computer viruses perform a lot of destructive actions on computers and networks and have caused organizations to lose millions of dollars. Spyware, on the other hand, is a computer program that clandestinely obtains information from a user's computer without the user's knowledge or consent. Keyloggers are an example of spyware.

Keyloggers have become popular and are an easy way to hack or illegally access computers. Keyloggers log computer users' input; oftentimes it is sensitive information like passwords, bank account numbers, and so on.

Internet cafés or areas that provide public Internet access are not the only places where security is negligent. Even prominent financial institutions have security issues that they may not be aware of. For example, most employees in Liberia use instant messengers, social networks and other media that have the potential to inject malware in their system. And system administrators may not be aware of this because they lack the capacity or tools that allow them take the necessary security measures needed to avoid a security breach. The plain fact is that not everyone thinks about security in the same way, although everyone gets concerned when they hear about a new virus on the Internet or when someone has hacked into a system stealing money, valuable information or simply destroying critical files. These acts can cause an organization to lose millions of dollars. And what's more interesting is that not every organization applies the same culture of security awareness that is applied in Western or developed countries. This is alarming, because hackers can penetrate any system of their choice for any reason. And there is a possibility that they will turn to areas that have poor security mechanisms when they discover that their targets in the Western world have become more difficult to attack.

Internet cafés do not enjoy the physical and logical security controls that are available within other institutions. It is therefore much harder to ensure that sensitive business information remains safe at Internet cafés or any area that offers public Internet access. Their attempt to institute security measures for their customers/users using technology may not be adequate, as there are still risks of shoulder surfing, dumpster diving, or even security cameras from within, observing keystrokes and computer screens. This is why providing security for networks should not be left only to the technology but should also include physical supervision by humans.

Governments, corporations, and organizations that have servers running critical applications or storing critical data should have strict policies governing the use of their systems. Mobile and third-party devices—external storage devices like the flash/jump drive used in public places—should not be allowed on government-owned or other institutions' networks. A government official *must* be prohibited from taking government-owned laptops to an Internet café or any place that provides public Internet access. The computers and laptops belonging to the government of Liberia *must* include the requirements for physical protection, access controls, encryption, backups, and virus protection. They should also include rules on connecting shared or mobile devices to corporate networks anywhere other than areas sanctioned by the government.

Internet computing requires conscientiousness and caution, on the part of both the user and systems administrator. To be conscientious and cautious, everyone needs to know how computers get exposed to security threats. Below is a brief list of the few ways computers can get infected by malware.

- Through portable or other media that users can exchange
- Opening e-mail attachments that are viruses
- Not running the latest computer, software, and antivirus updates
- Pirating software, music, and movies
- Not using antivirus and spyware scanners
- Downloading infected software

There are several different ways a computer can become infected with spyware, viruses, and other malware. But it is basically the user's actions that cause security issues.

A Few Steps To Take To Prevent Computer Systems From Being Compromised

The first thing that needs to be done is to ensure that there are policies in place that regulate Internet cafés and their operations; this could be done by the government's designated regulator. Then those who run Internet cafés have to be given the necessary ICT security training in order to establish some sort of control. The next step is to install and keep up-to-date *all* necessary security mechanisms that can protect the Internet café. If monetary resources are a challenge, then I suggest open source software that has several free and formidable anti-malware solutions. Install a firewall and an antivirus system that also has anti-SPAM capabilities. Also install a real-time protection program (RP), and again, make sure your security system is frequently updated; this includes training as well. Another option that is also popular these days is the use of software that can revert the system back to its original state. DeepFreeze from Faronics (www.Faronics.Com) is software that does a good job at reverting a system back to its original state, but there are others that do the same. Even though all that I have suggested are security measures, user training is paramount and cannot be overemphasized.

We are in an age of technology and the Internet; computer users need to be aware of existing security threats in order to understand the challenges that they may face. Those who run Internet cafés must take responsibility for this initiative. Owners of Internet cafés are, to some extent, liable for their users' security. Customers are paying for a service and therefore, they expect that their Internet or computing experience is safe. This is not an attempt to scare or deter Liberians from using Internet cafés and/or ICT but rather an attempt to raise awareness to prevent fraudulent acts or any form of cyber attack. It is imperative that cybersecurity is taken seriously if ICT is going to be the enabler of economic development in Liberia.

CHAPTER 23

THE NEED FOR PROACTIVE CYBERSECURITY MEASURES IN THE WAKE OF THE PROLIFERATION OF FINANCIAL INSTITUTIONS

The weekend of July 4, 2009, witnessed a blitzkrieg of cyber attacks on several Web sites of critical institutions in the United States and South Korea. These attacks temporarily crippled important organizations and some departments of government in those countries. The U.S. Treasury Department, Secret Service, Federal Trade Commission, Transportation Department, New York Stock Exchange, *New York Times*, and a few other private and critical institutions were reported to have been affected. In South Korea, the presidential Blue House and the Defense Ministry, along with a plethora of private institutions' web sites, were also affected. While both countries have advanced cybersecurity systems, these attack and previous attacks clearly illustrate the vulnerabilities of institutions and countries. These incidents experienced by the United States and South Korea should serve as a wake-up call to Liberia and other countries about the need to implement security measures on business and government networks. In this chapter, I will discuss cybersecurity in a new Liberia as it relates to financial institutions and other new businesses. I discuss a few known vulnerabilities and provide some countermeasures that could be used to prevent potential cyber attacks.

As financial institutions increase in Liberia so too has their use of modern technologies to enhance business strategies, facilitate operations, improve customer service, and gain competitive advantage. In these kinds of situations, the issue of security not only becomes imperative but also plays a critical role in their ability to succeed. Cyber theft, cyber scams (including "Nigeria e-mail scams"), cyber vandalism and "cyber espionage" are all a part of the new cyber war that is being waged globally on institutions. The culprit, be it a sixteen-year-old kid launching an attack from his parents' garage, a disgruntled employee, an experienced or novice hacker, a thief, a spy, or an entire country, sets out to achieve a goal. The result of that goal oftentimes can be colossally devastating.

Cyber attacks are often perpetrated by unscrupulous individuals attempting to penetrate computer systems/networks for several reasons, including: financial gains, espionage, and political purposes. Whatever the motive, a cyber attack presents a threat to a country's national security. The U.S. Department of Homeland Security reported that in 2008, the number of known breaches to its computer systems totaled 5,499, which is an increase from the 3,928 reported in 2007 (Ramsey 2009). Is it not interesting that the Web site of the Department of Homeland Security of the U.S. government could be breached that many times? One has to wonder what would stop cyber criminals from routing their attacks toward underdeveloped countries that have little or no "cyber defense" systems that rival America.

As previously mentioned, despite a gradually improving ICT sector, Liberia continues to experience a proliferation of financial institutions that have aligned technology with their business strategies to achieve their goals. While new investments have and continue to provide several opportunities for Liberia, the threats that accompany those opportunities must not be ignored. It is imperative that Liberia be proactive and institute effective computer security plans and processes that address the three major areas briefly explained below:

- ***Physical security***—This includes the protection of institutions' assets and information from physical access by unauthorized persons to include unauthorized access to data, servers, and computers by company employees.

- ***Operational security***—This involves the processes done within the institution, including the use of computers, networks, communications systems, and management of information.

- ***Policies and management***—This is the set of rules and procedures that govern an organization and *must* be supported by managers to ensure full implementation of its security plan.

Addressing the above three areas will prevent the possibility of a potential cyber attack.

In today's workplace, many of the technology resources we use to facilitate our jobs create more vulnerabilities than we had envisioned, especially if they were not installed properly; hence the need for conscientious systems or security administrators who have the ability to educate users, keep apprised of developments in ICT, maintain an up-to-date security platform, enforce security policies uncompromisingly, and be vigilant at all times. In the next section, I explore a common communication tool that is gaining popularity in institutions around the world: instant messaging!

Instant Messaging And Security

Communication is not only crucial, but it is also imperative and is an integral a part of one's daily routine. Many institutions use a wide variety of media for communication: telephone, e-mails, instant messaging, etc. While these are all effective communications tools, instant messaging has grown and is being widely assimilated into society as a cost-effective option. Since its evolution, instant messaging has now become the third-most-widely used form of communication in succession to the cellular phone and e-mail. It has changed the communications landscape dramatically and has also been incorporated into business environments. It has lots of advantages over e-mail, such as instant responses and file transfer capabilities, etc.

Yahoo! Messenger is a widely used instant messenger program that has further broken the communication barrier—initially brought about by the digital divide—between developed and underdeveloped nations. But the nature of Yahoo! Messenger and many other instant messengers enables users to bypass security gateways and other cyber-defense mechanisms, allowing individuals on the other side of the "firewall" to gain unauthorized access to a network. With this ability, viruses and other malware are able to gain entry to networks that are otherwise secured.

Some instant messengers, including Yahoo! Messenger, for example, allow the use of Webcams and file transfers; this also allows employees or individuals to manipulate or steal confidential files and transfer them to competitors without notice. Instant messengers also have buffer overflow and serve as the conduit for malware dissemination and denial of service (DoS) attacks. Their content cannot be filtered or archived and they do not leave any trails for audit, thus presenting legal and compliance problems for firms. Despite the threats they pose to mission-critical data and networks, instant messengers do provide a great deal of convenience and are cost effective; of course this depends on how well they are implemented and controlled.

In Liberia, the proliferation of banking institutions inherently creates more opportunities. Unfortunately, however, these opportunities also come with threats that present serious risks if not addressed properly and effectively. To gain competitive edge, several strategic resources must be put in place, some of which are innately complex and accompanied by a myriad of security vulnerabilities. Computers, laptops, networks, etc., which are part of modern business strategies, face several sophisticated attacks on a daily basis and must be deployed with countermeasures. Below I list some of vulnerabilities that I believe individuals working in computing environments should be aware of.

- Fake advertisements/malvertisements that redirect people to malicious sites; for example, search engine redirection

- Vulnerabilities in servers such as open network ports—e-mail (port 25), Web (port 80), or other ports. Ports 25 (e-mail) and 80 (Web) are frequently targeted.

- Web browsers (Internet Explorer, Mozilla, Safari, etc.), ActiveX controls, browser-plug-ins, multimedia, and other third-party applications with any vulnerability will automatically create a loophole, thereby opening your server or computer to an attack.

- Failure on the part of IT managers or security managers to update computers with latest securities patches
- Default operating systems installation; deploying operating system without configurations
- Vulnerable CGI programs
- Accounts with weak or no passwords
- Using computers with little or no education about computer security gives malware writers a larger audience and more victims. This is because these users only focus on the content rather than the potential harm of some of the applications or advertisements with which they interact.

Some countermeasures include:

- Installing intrusion prevention systems (IPS) and intrusion detection systems (IDS); these systems help prevent and detect intrusion and can be very effective if the proper algorithms are set
- Prohibiting the use of free and non-authorized instant messengers on computers with "mission-critical" files or applications
- Installing layered security to ensure confidentiality and integrity of data
- Requiring users to have strong passwords using strings; for example, D@rReN*2o09 is better than darren2009.
- Controlling the flow of data and Internet usage
- Encryption of critical data
- *Education, education, education*—If users are informed of threats and vulnerabilities, they will apply and support security measures, especially if they know that not doing so will result in termination of services.

ICT managers need to be ready and prepared to effectively respond to crisis situations. Below are a few disaster-recovery procedures (DRP) that ICT managers should consider.

- Establish a strong backup system and storage system at a secured and remote location. For example, if you have an office on Broad Street, make sure that your critical data are backed up and stored offsite; somewhere in Kakata or Zwedru, or maybe at your satellite or regional offices.
- Check how long it will take you to restore those backups and your entire system. Customers should not have to wait too long after an incident. Your prompt response to a disastrous situation ensures company concern and reliability.
- Identify critical systems.
- Test your system and backup restorations and verify them as often as you can.
- Prepare to restore your system as soon as possible in the event that there's a disaster.
- Devise a disaster-response team and policy for after-hours issues.

The above listed procedures should form part of your disaster-recovery plan (DRP), which should include other exercises not mentioned here. In a digital world, it is imperative that an IT administrator/manager maintains a comprehensive disaster-recovery plan.

We should learn from the recent attacks on the United States and South Korea that was perpetrated using a "botnet" virus composed of about fifty thousand infected computers. These computers were in a denial of service (DoS) attack. These kinds of attacks are a nuisance to ICT departments/environments but again can be countered easily provided individuals and companies perform due diligence. Countermeasures should involve all stakeholders, including Internet service providers especially, since they are used as conduits. ISPs are capable of filtering outbound traffic and identifying unknown or spoofed source packets attempting penetration and then dropping them, thereby precluding a possible attack.

Finally, cyber criminals periodically change their modus operandi. This means that we, as ICT managers, must be ahead of the game. Traditional antivirus software and other security solutions are ill-equipped to handle the dynamic attacks of cyber criminals. Hence there is a need for institutions to frequently educate their users about cybersecurity and to keep their computer systems/networks up-to-date with security patches. We need to always remember that business strategies and processes change, and cyber culprits parallel their modus operandi with those changes. Therefore, we must adapt to these changes if we wish to remain competitive. In Liberia, heightened awareness and the need to be proactive are key to advancing technology seamlessly and securely.

PART VI

HEALTH

CHAPTER 24

TELEMEDICINE—PROVIDING HEALTHCARE BEYOND BOUNDARIES

This chapter discusses telemedicine, the fusion of ICT and healthcare, and how it will increase access to quality medical care with special emphasis placed on the rural areas. Telemedicine applications can drastically improve Liberia's healthcare sector, greatly improving the quality of life and life expectancy of its citizens. However, for the benefits of telemedicine to be realized and in order to provide viable healthcare there are many aspects of this sector that require a shift in paradigm.

In recent years the availability of health workers in and around the capital city of Liberia has and continues to grow, but health workers—especially nurses who serve the rural population—are isolated from specialist support, up-to-date information, and opportunities to exchange experiences with colleagues. As Liberia's partners and stakeholders (NGOs and other organizations) make available their resources (helicopters, vehicles, etc.), the unavailability of much-needed technology to facilitate healthcare in rural areas presents a challenge for nurses and other healthcare workers. With telemedicine, many of these challenges can be alleviated and many lives saved because information will be shared effectively and promptly.

Telemedicine is the use of telecommunication technologies to provide healthcare services over long distances and can be used to improve healthcare as well as enhance economic development in a developing country like Liberia. It allows physicians to provide consultations and collect and share laboratory data and medical records to provide medical care. The introduction and implementation of telemedicine in Liberia will change the course of healthcare delivery. Telemedicine will enable doctors at John F. Kennedy hospital to be able to diagnose and treat patients in real time, in remote areas like Weasua or Yekepa, from their offices in Monrovia or even from the United States. It is an ideal counter-response to threats of national disaster because it ensures that real-time medical care is available to potential victims despite geographical locations.

With Liberia at such a critical juncture in the recovery process and with a considerable number of its population living in rural areas, the implementation of a national telemedicine program will be positioned to address the following issues:

- The scarcity of specialists
- The lack of other healthcare professionals in rural areas
- The lack of medical equipment in rural areas
- The lack of good roads and transportation
- The lack of effective media for communications
- The lack of proper health education for the population in rural areas

Telemedicine allows doctors to perform several tasks remotely and conveniently; tasks that in the past required patients to physically visit hospitals, costing them time and money for transportation.

Healthcare for pregnant women and their babies will improve significantly, because there will be an increase in prenatal care and doctors will be able to perform fetal monitoring remotely and consult health workers to examine "prospective" mothers in order to keep apprised of their condition.

Telemedicine can make a difference in how healthcare to the poor is delivered. Statistically, rural areas have been listed as those areas with a high population and the poorest of citizens. The implementation of a national telemedicine program that can reach the population in rural areas will greatly increase its life span.

I am confident that telemedicine will work well in Liberia because of the global advancements in technology and telecommunications, and progress being made in Liberia, especially in the area of mobile telecommunications. Mobile telecommunications and technologies have changed the way Liberians communicate. GSM, CDMA, and other emerging mobile technologies continue to bring affordable and easy mobile phone access to areas previously known to be devoid of communication apparatuses. The advent of broadband connectivity will enable video conferencing, allowing the use of interactive media that will permit doctors and healthcare workers to provide quality healthcare to all Liberians.

Another issue that needs to be addressed is the fundamental and continued education of health workers to enable them to provide the best care to their patients. Nurses, who along with teachers are my most revered professionals, must be given the best education and tools to keep up-to-date on the changes that consistently occur in the medical field. New technologies continue to change the way things are done; hence they need to have access to current information relating to their profession. Also, apart from having nurses attend nursing schools or colleges, Liberia's health system should provide technologies that give them access to the Internet where they can gain a wealth of up-to-date information and resources to better themselves as well as connect, collaborate, and communicate with other healthcare workers in the global community. Below are a few suggested approaches that Liberian healthcare authorities should consider implementing:

- The development of a healthcare information system that will be named Health Information and Technology Communications Hub (HITCH). HITCH will link *all* medical centers in Liberia and provide healthcare professionals with up-to-date information on medical treatments, access to medical files, referral system to specialists, and continuing education. From the consumer perspective, HITCH will provide directions to health centers, basic information about common medical conditions, preventative healthcare recommendations and behaviors to avoid, and appointment scheduling.

- The development of an e-health initiative that will enhance the quality and efficiency of healthcare in Liberia through ICT, particularly through the use of HITCH
- Allow nurses to travel to Western countries to attend training programs and bring back best practices of their profession and spread them across the healthcare spectrum
- Provide nurses with PDAs (personal data assistants) containing up-to-date basic reference materials as part of their practice and continuing medical education
- Provide all necessary technologies to healthcare professionals so that they are informed about trends in healthcare and the practical impact, positive or negative, that new standards, technologies, and products will have
- Provide access to e-mail, chat, and instant messaging to healthcare workers. These tools can be used on PDAs (i.e., Palm Pilot). E-mail, chat, instant messaging, and now video conferencing are very economical solutions to support healthcare in remote areas because of their store-and-forward functionality. All of these tools are very user friendly and are available in the open source community for free, or they can be obtained as COTS (commercial off-the-shelf) solutions. They allow the sending of e-mail attachments such as image files, thereby permitting a form of low-cost telemedicine. Video conferencing can also be done in a low-cost way by using the appropriate application with cheap hardware (Webcam). Skype is an excellent tool that is ideal for healthcare workers. Its videoconferencing features are great for communications between doctors in the main city and nurses in the rural areas.
- Extend satellite capacity (VSAT) if not broadband capacity to villages. This has to be done through or in collaboration with LIBTELCO or any available mobile operator or ISP.

- Purchase vehicles that can serve as mobile healthcare clinics and are equipped with telemedicine technologies to be used primarily in rural areas allowing treatment of patients and training of rural healthcare workers.

- Create a tele-health medium to improve healthcare via tele-nursing, teleconsultation, telemonitoring etc. Tele-health can be two doctors using information technology and communication to provide healthcare to a patient or a patient receiving healthcare from a doctor using ICT.

- Implement a general and anonymous health helpline with specific areas of concern: HIV/AIDS, STDs (Sexually Transmitted Diseases), tropical diseases, and other sections that reflect the medical concerns of Liberia. For example, an individual calls the general health helpline and has questions/concerns about HIV/AIDS. That individual will be transferred to that department, allowing them to seek more in-depth information and also report any illness or symptoms they are experiencing related to HIV or AIDS. The health helpline, which is a telephone service or chat service operated by a medical staff member of the Ministry of Health, will assist, and appropriately direct individuals to one of the specific areas of medical concern. The staff member will provide confidential counsel and referral to medical facilities if an individual chooses to report an illness or incident of illness that may require medical attention. If a medium for anonymous reporting is established, more medical cases will be reported, thereby allowing comprehensive information gathering about the health of Liberia. In addition, there will be a decrease in the number of Liberians dying from preventable and treatable diseases.

- Installing professional development centers in major hospitals in every county and requiring *all* county health employees to attend regular in-house training. Each hospital's professional development center will be equipped with at least twenty computers, a projection device, Internet access, and health-related software. Internet connectivity will also be needed. Trainers will be products of a "Train the Trainer" program.

Now, while I advocate the use of ICT to improve our healthcare system, we should not be tempted to directly emulate the Western healthcare system, which is definitely different than ours. Western healthcare systems have standards and procedures that might not be applicable to developing countries like Liberia. And unless Liberians understand the technological and cultural readiness of their country and its healthcare practitioners, embarking on this initiative will be fruitless and may even cause more problems.

ICTs have unleashed new opportunities for the delivery of health services. Despite these new opportunities, the Liberian Ministry of Health is yet to implement any of them. Hence, the Liberian healthcare sector is still faced with several challenges in meeting the healthcare requirements of residents, especially those in the rural areas of the country. Some of the challenges are little to no access to emergency treatment, inadequate information dissemination, inadequate infrastructure, etc. With so many inadequacies coupled with increased demand for improved healthcare delivery, telemedicine tend to be the ideal solution to address the challenges of healthcare delivery in Liberia.

Part VII

Agriculture

CHAPTER 25

E-AGRICULTURE AND M-AGRICULTURE—NEW AND BETTER APPROACHES TO AGRICULTURE IN LIBERIA

In this chapter I shall discuss revamping the current "farm-to-market" process in Liberia's agriculture sector, which unfortunately puts farmers on the lower side of the pendulum; that is, not getting their products exposed to a larger population for competitive prices.

In a letter injected into Liberia's Poverty Reduction Strategy document, President Ellen Johnson Sirleaf wrote, "The PRS lays groundwork for making sure that the child's parents have a fine road to carry their goods to market and can participate in a local government that is vested with increasing responsibility and resources" (International Monetary Fund 2008). This statement was made in reference to expectations made by stakeholders from rural Liberia during the preparation of the Poverty Reduction Strategy document. The president is saying that a smooth path to the market will allow farmers time to tackle other responsibilities. But what if we create a situation in which the farmer does not have to leave her farm? What if we implement e-agriculture and m-agriculture to take the burden of traveling to the market off the farmer?

In the subsequent paragraphs I will discuss e-agriculture (or electronic agriculture) and briefly m-agriculture (or mobile agriculture).

I believe that with the e-agriculture approach, the farmer may not have to spend days traveling to the market. Instead she will have her goods picked up by a courier and taken to the market, thereby giving her time to engage in other activities that can further increase her production.

The e-agriculture approach will bring improvement in rural and agricultural development initiatives through the exchange of information. I use ICT in this chapter to shift from the old, tedious, less-profitable method of agriculture to a more modernized, convenient, and potentially profitable approach. E-agriculture combines three separate technologies to accomplish its goals. These combined technologies include communications technology, computer technology, and information management. When combined, these technologies provide information sharing, knowledge, data management, and fair transactions.

The approach involves the following: installation of an e-commerce Web site/portal, links to financial institutions, installation of knowledge centers in counties (cities) with access to database/backend servers, equipping village head(s) with mobile phone(s), since they may not have computers, and so on. It will require the involvement of all stakeholders, including county officials, the Ministry of Internal Affairs, chiefs, ICT officials, the Ministry of Agriculture (MOA), banks, district trainers and educators, farmers, couriers, Liberia Produce Marketing Corporation (LPMC), the Ministry of Commerce, and so on.

How It Works

Farmers or town chiefs will use mobile phones to communicate with a "hub" in each county. This hub, which will be erected by the Ministry of Agriculture (MOA) or LPMC will serve as a conduit between the farmer and the market. When the hub receives the information requiring the types and amount of products available for market, it enters the data into the system, which is connected to the main databases of the Ministry of Agriculture or LPMC.

The information is then posted on the MOA's or LPMC's e-commerce Web portal, where customers can go and purchase items using their banking credentials, etc. Delivery of the product would be an additional service provided by the Ministry of Agriculture, the Liberia Produce Marketing Corporation or an independent courier. The system will allow the recording and uploading of photos and videos of products, process description, route information, and every type of information that can create convenience and add value to the process. Information exchange in this setup will be accurate, concise, and complete. Since some goods might be perishable, information will be provided in real-time. The system will be user friendly and secure although easily accessible.

While there is a need to create national literacy, this initiative presents a solution for situations that may not require tremendous investments in education. Using mobile phones to communicate with a representative at the local hub or access point who will be bilingual will greatly break the communication barrier and the need for computer knowledge or experience in using an e-commerce Web site.

Mobile agriculture or m-agriculture works similar to e-agriculture. The only difference is that m-agriculture uses mobile devices, especially mobile phones, for transaction. The data from mobile phones goes through the same path as the data from e-agriculture. Since mobile phones have penetrated more pervasively into the rural sector, it is most likely to lead to a successful m-agriculture initiative.

This, of course, will not be the only medium used to improve our agriculture sector. Other information and communications technologies, such as radio and television broadcast in local languages or the installation of call centers, would also be included.

E-agriculture is not a concept I created. It was acknowledged at the World Summit on the Information Society in 2005 and is being embarked upon by the Food and Agriculture Organization. An e-agriculture implementation in Liberia will transform the agriculture sector and bring economic prosperity to those in the rural areas in particular and to Liberia as a country. Obviously, this entails some challenges, but I strongly believe that both e-agriculture and m-agriculture can be adopted in Liberia successfully. Liberia can begin and enhance this process by including e-agriculture in colleges' curricula and by using every available media to create awareness and literacy for all stakeholders.

Part VIII

Future

CHAPTER 26

ONLINE BROADCASTING: LEVELING THE PLAYING FIELD FOR INFORMATION DISSEMINATION

The Internet and its technologies have brought all sorts of new and exciting ways to do things. Since its advent, almost every aspect of society has been impacted. This includes the broadcasting industry. This chapter is based on one of my articles published in the *Liberian Daily Observer*. I figured I have done enough "disservice" to the broadcasting industry in this book by not saying much about it. Moreover, a book about a digital Liberia will not be complete without discussing the broadcasting industry, hence this chapter on online broadcasting.

There are a growing number of Internet radios and television stations—some with the dot.com; others with the dot.info URL. These Internet stations have embarked upon what we now call online broadcasting. Online broadcasting is not novel in a global sense but is so in a less-privileged country like Liberia. It seems to be penetrating the Liberia arena overwhelmingly in the Diaspora and gradually in Liberia. Online radio and TV stations are commandeering large audiences and are changing the way information is being disseminated to societies.

Online broadcasting is done through streaming over the Internet. In the case of online television, or IPTV or Internet protocol TV, videos are displayed through streams that are encoded as a series of Internet protocol (IP) packets (Anderson 2006). Basically, when you watch

TV over the Internet, you are watching IPTV. IPTV is the successor of satellite service, digital cable, and HDTV (high-density television). Online radio does the same, although it does so with audio instead of video.

I predict that the online broadcasting industry will continue to grow and that they will impact Liberia and the entire continent of Africa greatly, if they have not already done so. With cellular phones turning into smart phones and the proliferation of submarine fiber-optic cables in Africa, the impact of online broadcasting will be revolutionary. But this revolutionary impact brings forth several questions as it relates to Liberia: Will there be as many radio and TV stations as there are political parties? Will universities run their own station and allow mass communication students to gain practical knowledge? Will the Ministry of Education run an education broadcasting network? How will online radio impact Liberia? It is already impacting Liberians in the Diaspora! Will cell phones act as the radio of the future? I don't have all the answers to these questions, but below I give you some of the changes that online broadcasting will bring to Liberia and how I feel they can positively impact our recovering nation.

1. *New or struggling artists:* Online radio stations can and will have a great impact on the entertainment industry, especially for little-known artists. With the Internet and Web 2.0 features, a lot of Liberian artists will have greater opportunities to introduce their work to new audiences.

2. The Ministry of Education will be able to establish an online station that broadcasts educational programs to help lower the illiteracy rate in Liberia.

3. The University of Liberia and other colleges will be able to use online radio and television stations in mass communications program. These options can also be used to provide information for current students, prospective students, and alumni.

4. Political campaigns or political agendas will be broadcast to the public at lower costs. Recordings of candidates' speeches,

analysis, and commentaries will be the focus of these online broadcasting stations.

5. New and old businesses will be able to increase their visibility and revenue through this new marketing avenue. Businesses will have another medium for cheap advertisement, allowing them to reach existing and new customers.

6. The Health Ministry will be able to operate an online radio station to provide information to Liberians about their health and available options.

This list is not exhaustive. There are more uses for online broadcasting in Liberia. The bottom line is that today, almost anyone with a message can broadcast to the world for little or nothing. Never before have there been so many opportunities to empower a plethora of people with a plethora of communication possibilities. And the possibilities abound.

There are several approaches to broadcasting online that a person can choose depending on his or her goals. If you want to make a profit or simply put out a message, modern technology now provides several options. It doesn't require a lot of skills or capital. That's the amazing part! There are even options for a novice that require very little technical knowledge. All that a person needs to do is create some files in recognizable and compatible formats, upload them, make a few configurations, and they would be ready to reach a global audience.

Liberians should look to online radio stations to uncover their (online stations) far-reaching implications on society and to understand their legal and regulatory details. There may be confusing legal and regulatory obstacles that might have to be addressed in order for online broadcasting to succeed in Liberia. I mean, how do you regulate a radio station that's based on the Internet operating from the United States or Guinea? In the United States, there had been debates about online radio stations. An example of this is the anti-performance tax resolution against the recording industry of the U.S.A. that Senators Blanche Lincoln (D-Arkansas) and John Barrasso (R-Wyoming), along with Representatives Gene Green (D-Texas) and Michael Conaway (R-Texas) and several congressional members worked on some time ago (National Association of Broadcasters n.d.).

With cloud computing, broadband, and other advances in technology, online broadcasting will continue to become easier, cheaper, and ubiquitous. The Liberia Broadcasting System and other radio stations that spend thousands running their stations through the traditional setup should start looking into this matter to figure out their next approach to maintaining competitive advantage.

CHAPTER 27

WHY LIBERIA MUST TURN TO OPEN SOURCE SOFTWARE (OSS)

In Liberia, Microsoft Windows and Microsoft Office are the de facto standards for personal, educational, and business computing software. However, these products are costly to large corporations in a country like the United States and therefore must be a financial albatross for a country like Liberia. The mere purchase of proprietary software is costly, but it is even costlier when calculating total cost of ownership (TCO). In view of this, I am of the conviction that if Liberia continues to limit itself to the use of proprietary and commercial off-the-shelf (COTS) software like Microsoft Windows, it will not only limit its capacity but will also limit its chances for innovation and creativity. And this would adversely impact the benefits that Liberia can garner from information technology for its economic recovery initiative.

In this chapter I discuss why Liberia must turn to open source software to create a knowledge society. I discuss the impact of open source software on the world today and how other African countries are leveraging it to improve their economies.

In November of 2008, I wrote the initial technology implementation plan for the B. W. Harris School Smart Tech project. In that plan, I had included the installation of a LAMP (Linux, Apache, MySQL and Perl/ PHP) server, which I had planned to use to set up an Intranet. Before I did so, a colleague of mine who is also an ICT professional residing in Liberia was quick to discourage from installing the Linux server. He claimed that "there may be very few" ICT professionals in Liberia with knowledge of the Linux operating system. Initially, I differed with him because I could not imagine why Liberia, a country struggling to build an ICT sector that will help bridge the divide, would not be willing to take advantage of free and open source software when other African countries have already done so. I would later learn that despite the country's efforts to enhance ICT penetration, the government has not done much to encourage the use of open source software. The most it has done in terms of encouraging the use of open source software is the mentioning of open source software in its national ICT policy document.

Unfortunately, though, during my visit to Liberia that year (2008) and the subsequent year (2009) I discovered that my colleague's remarks were indeed true and that most of the computers in Liberia ran on a Windows platform. Because of Liberia's dependence on the Windows platform, many ICT professionals, I presumed, would not be familiar with the Linux operating system, which is an open source software.

My presumption was confirmed when I had a meeting with fifteen ICT professionals in December of 2009. The meeting was an effort to recruit members for the Liberian Open Source Initiative (LIBOSI), an organization which I had started to enhance the penetration of open source software in the Liberian society. In that meeting I asked the participants if anyone knew about Linux and was told that they had heard of it but had not seen or used it before. That revelation caused me to ask myself, "Will Liberia ever bridge the proverbial digital divide if it relies on proprietary software?" Not really! Can Liberia ever be a nation of ICT innovation? *Possibly*! But only if Liberia embraces and engages open source software will it succeed in bridging the digital divide and become a nation of innovation.

So, what is open source software and what has it done for economies? Open source software is software that is freely licensed to users to study, change, and improve its design through the availability of its source code. OSS, as it is referred to, has gained momentum over the years in developed, developing, and underdeveloped countries. Open source is "free" as in freedom!

Since it became a formalized model for sharing software, open source software has been a force to deal with in the business world (Hahn 2002). It introduced a relatively new paradigm in the software industry, which has since led to it being a tough competitor against proprietary software. The philosophy behind open source software has made it ideal to use in environments that are financially strangulated.

Open source software provides a multitude of solutions that can be used for computing and other purposes. Linux, Apache, OpenOffice, Firefox, MySQL, Java, PHP, etc., are a few of the plethora of free software options that the open source community offers. Proprietary software like Microsoft Windows, Internet Information Server (IIS), Microsoft SQL Server, and Microsoft.Net are few software options with prohibitive cost that countries like Liberia cannot afford.

The growth of open source software in modern economies has been a result of its licensing model and not the programming language used, the operating environment, or the application domain (Tiemann 2009). Because the licensing allows anyone to make changes to the code, open source software allows innovation to permeate societies, leading to very low-costs products. "Open source software has changed conventional assumptions of software in the industry," argues Michael Tiemann, vice president of the Open Source Initiative in his essay written on open source software (2009). Also, according to a research done by Deshpande and Riehle (2009), between 1995 and 2006 one billion lines of source code were written in open source software in every 12.5 months. When these lines of code are calculated in terms of proprietary software, it would cost millions of dollars to produce.

The Linux operating system, which competes with Microsoft Windows, runs the majority of the servers on the Internet, and is a product of the open source community. Ubuntu Linux, the most popular distribution of Linux, is now being installed on desktop computers being manufactured and shipped by well-known computer manufacturers including Dell Computers.

The continent of Africa, home of many developing nations, has adopted open source software to provide many of its computing needs. This adoption of open source software has led to several developments in Africa, many of them concentrated in southern and eastern Africa. Furthermore, the Free Software and Open Source Foundation for

Africa (FOSSFA), was launched on February 21, 2003, in Geneva. This organization advocated and continues to advocate the use of open source software in several African countries, including South Africa, Tanzania, Uganda, Kenya, Egypt, Ghana, Benin, Mali, and so on. Today, open source software has become so popular in African countries that some countries have begun developing software in their native languages. In Tanzania, where Swahili is spoken by an estimated one hundred million people, a software called Jambo OpenOffice has been developed. This software, which is based on OpenOffice and runs on the Linux operating system, is also translated into Swahili. Jambo OpenOffice has functionalities similar to those of Microsoft Office and is available for free (OpenOffice.org 2004).

The government of Liberia, whose IT platform is Windows-based, needs to start exploring open source software to minimize its IT spending and encourage creativity and innovation. Open source has positioned itself as a genuine and ideal approach to finding IT solutions for financially challenged countries than proprietary software. Furthermore, cloud computing, which includes several software applications that are open source, has again kindled a renewed interest in open source products. Several cloud infrastructures, such as Google, Amazon, Rackspace, and other smaller startups, have relied heavily on open source software, while key open source providers are targeting their efforts toward the cloud.

The adoption of open source offers great potential, especially at a time when cloud computing has opened the way to a new generation of computing. Globally, businesses and governments will gravitate toward open source software because the open source community has made overwhelming progress in making its software consumer-ready and user-friendly. If it continues to move at its current pace and focuses on becoming customer-centric, it will most likely be the toughest competition that proprietary software ever had.

The Liberian National ICT Policy document discussed earlier addressed open source software briefly; however, more needs to be included in terms of the government's policy and involvement with regard to open source software. The government of Liberia will also need to provide incentives to businesses and all other stakeholders, perhaps by giving tax breaks or subsidizing open source software use and research.

Universities and technical colleges will need to set up computer science programs that will perform research work in open source technologies. Most of all, the government should set the precedent by using open source software and allow a pervasive penetration of ICT in all regions of the country. It should encourage the use of open source software in schools and institutions of learning to save on the cost of licenses and to promote creativity.

Besides eliminating the cost of licenses, most open source software can run on obsolete hardware since they do not require as many resources as their proprietary competitors do. In addition, the Linux operating system comes with more than three hundred distributions, most of which come loaded with OpenOffice or some sort of productivity suite. This can be used in schools as an effort to integrate technology into lessons at low costs. Schools can also opt to use Ubuntu Linux, or the educational version Edubuntu which is ideal for K–12 schools.

Open source software guarantees Liberians the opportunity to develop local programs built by Liberians for use in Liberia and perhaps on other parts of the continent. The time has come for the Liberian government to begin initiatives that will introduce open source into schools, where young people can gain a fundamental knowledge of open source and learn to create, use, maintain, modify and improve computer software.

Finally, as I briefly mentioned above, during my trip to Liberia in December of 2009 and after meeting with ICT professionals mentioned above, I decided to form the Liberian Open Source Initiative (LIBOSI). LIBOSI is a not-for-profit organization founded to promote and support the use of open source technologies in Liberia for economic development. Its aim is to use open source technologies to build ICT capacity. Before departing Liberia, I urged the group of ICT professionals that I had previously met with to continue what I began to ensure that LIBOSI becomes a stakeholder in the Liberian ICT sector and is recognized internationally. LIBOSI has been growing gradually and aims to be a part of the change that will occur in a digital Liberia.

Chapter 28

Green IT And Virtualization

I find it very difficult to work with only one computer or one monitor because I am the quintessential "multitasker." So, befittingly, one would assume that I would have an exorbitant energy bill. A while ago, I would have agreed with that assumption. However, after switching from a multiserver and multinode computing environment to a consolidated and virtualized environment, I found myself running a converged home IT environment. This change lowered my energy bill by about 30 percent and has led to what I call my personal "green data center."

ICT has become critical to every sector of our societies—firms, governments, households, etc. Without ICT, a business or organization may not be able to effectively compete in the global market. But this dependence on ICT has also increased our dependence on energy, which has led to our dependence on and "addition" of fossil fuels. In Liberia, electric energy is expensive; hence the need for alternative solutions that will minimize, if not eradicate, Liberia's total dependence on it. This means we should begin consuming electrical energy responsibly and explore different options for electrical energy and for the preservation of the environment.

In this chapter, I discuss Green IT and how it can reduce the cost of running IT environments in Liberia. I added virtualization to this chapter because I believe that it is critical in a "Green IT" environment, although it can also be addressed in the context of cost saving.

An archetypical enterprise IT data center includes data, people, policies, servers, hardware, networks, and storage, combined together in an effective and manageable manner. All of these individual components must work together in cohesion to provide services. Several components are needed in this environment: storage area networks/network storage devices, hypervisors running on physical servers, power backups, physical switches, virtual appliances, application-level proxies, Web servers and databases, and more. All of the above mentioned components require some sort of energy to function. In a country like Liberia where energy is scarce and costly, this will drive up the electric bills of government, businesses, individuals, and organizations that run IT data centers. Worst yet, it will increase Liberia's production of carbon monoxide, which greatly pollutes the environment. So how does Liberia deal with this kind of problem? One of the options would be to "go green." And "going green" entails a lot of things including running IT departments in a "green" way to reduce consumption of electricity and minimize the problems of environmental pollution. Basically, this means "Green IT." So what then is "Green IT"?

WHAT IS GREEN IT?

Green IT is a relatively new concept in the IT industry that basically involves the efficient use of energy through the installation of energy efficient devices, in addition to the consolidation and virtualization of IT equipment in the data center. A green data center is an environment that has a multitude of servers, networking devices, lighting, software, people, and processes; where energy efficiency and absolute minimum environmental impact are of the most importance. It is a place where data is stored, managed, and disseminated through efficient usage of energy. Constructing a Green IT data center involves more challenges than installing a traditional data center.

John Lamb, author of the book *Greening IT: How Companies Can Make a Difference for the Environment*, notes that Green IT provides cost savings for IT organizations. He points out that IT organizations have a scheduled three-to-five year refresh plan that allows them to get technology that brings them up to date with new technology. The refresh policy, according to Lamb, "provides a greater opportunity for a company to buy new energy-efficient equipment and implement server and data storage virtualization" (Lamb 2009). This poses another question: how does this apply to Liberia, which does not have large data centers?

Green IT In Liberia

Liberia is a country working toward development, and the imminent connection to the ACE submarine fiber-optic cable will allow companies to build data centers. Green IT will have significant impact on organizations in Liberia because it will provide tremendous cost savings and enhance the country's energy sector.

Green IT is not just about energy or cost savings; it's also about operational efficiency and maintaining a better IT environment. Everyone in the IT industry in Liberia must at the very least understand the implications of mass consumption of power, especially when there are limited amounts available. Awareness of IT's consumption of power leads to more efficient processes, reduced costs, and a "greener" environment. IT professionals in Liberia need to start thinking "green" since they play a more crucial role in their places of employment.

It's important to mention that the majority of the countries in Africa have unreliable power grids. Liberia is certainly no exception, as evidenced by the proliferation of privately owned diesel generators that run several hours a day and up to seven days a week. The use of these energy sources greatly increases Liberia's carbon footprint; but what other choices does Liberia have? I am not particularly familiar with alternative energy, but I believe that Liberia must begin exploring alternative energy to replace its current source of electrical energy. In Liberia, electricity has never been a national "luxury," but with alternative energy garnered in a green way, there might be a chance to electrify the entire country.

To go "green," IT departments in Liberia must be ready to do several things: ensure low-emission building materials are used during new construction processes, recycle waste properly, and use alternative energy technologies. Leaders and managers within organizations will need to encourage the use of "green" approaches. Leaders, especially those in the government sector, must be active in working to maintain Green IT environments. They must be able to regulate the use of energy as well as provide education to stakeholders about it.

The Liberian government should begin exploring innovative ways to provide energy to its citizens. This can be done either liberalizing the energy sector or by investing in research that will subsequently produce alternative energy.

Innovative energy has been the focus of many discussions in recent times as a way of enhancing data centers. Mauritius is an African country that has been identified as a potential IT hub that will connect Africa, Asia, and the Middle East because of its geographical location and ability to use alternative energy to run data centers. The use of the ocean—which surrounds the country—to provide a source of energy for data centers was contemplated as a means of cutting the cost of running them (data centers) as a "hub" (Miller 2009). Officials in Mauritius through its Eco-Park initiative plan to use sea water air conditioning, or SWA, to provide colder water originating from the deep water currents to cool off data centers on the island. The cool water coming from the deep currents will be piped back to the data center, meaning that there will be no need for chillers, which require a lot of power (Miller 2009).

Virtualization—The B. W. Harris School Desktop Virtualization Project

In 2008–2009, I, along with several others, "piloted" desktop virtualization at the B. W. Harris School in Monrovia. The desktop-virtualization project worked seamlessly, and the students were able to use the computers without noticing that they were running "virtual sessions" in a virtualized environment.

Virtualization, are "technologies designed to provide a layer of abstraction between computer hardware systems and the software running on them. By providing a logical view of computing resources, rather than a physical view, virtualization solutions make it possible to do a couple of very useful things: They can allow you, essentially, to trick your operating systems into thinking that a group of servers is a single pool of computing resources. And they can allow you to run multiple operating systems simultaneously on a single machine" (Waters n.d.). In layman's terms, virtualization is the creation of many virtual resources from one or more physical resources, in most cases, a server, network, or a storage system.

Virtualization is not as new a technology as it is thought to be. The concept dates back to 1959 when Christopher Strachey published his paper titled "Time sharing in large fast computers" at a UNESCO conference. Following that, IBM, VMware, and other companies and institutions continued working on virtualization, which is now being deployed all around the world (Singh 2004).

There are several types virtualization. The following are the most commonly found: server virtualization, desktop virtualization, application virtualization, network virtualization, storage virtualization, and operating system virtualization. Since desktop virtualization is one of the technologies that was implemented at B. W. Harris School in December of 2008, I shall delve deeper on that topic.

Desktop virtualization is defined in ICT circles as a technology that borrows from the thin-client model. The thin-client model is based on a client-server technology which means management is performed on the server and not on the desktop or client. It encapsulates applications from the underlying operating system upon which they are expected to run. In layman's terms, in desktop virtualization a computer that's running in a virtualized environment provides multiple instances of the same or different operating systems on other desktops through a device widely known as a terminal. Desktop virtualization gives the user a personal computer experience even though he or she is only using a monitor and a terminal.

There are many reasons why a business, an organization, or a school like B. W. Harris High would want to implement desktop virtualization. The primary reason is to maintain a centralized environment in which little is spent on installation, configuration, maintenance, and energy. Saving energy is done by replacing regular desktop PCs with virtualization terminals and/or software. Users' computers can be managed easily by using this approach.

Security is the second reason for running a virtualized environment. A third reason for running desktop virtualization is its support for legacy applications on modern operating systems that were not initially designed to run them. The final reason is cost. Desktop virtualization provides multiple accesses to desktops for a fraction of the cost of a physical computer.

From an economics standpoint, running virtual desktops is very cost-effective. In an educational environment, desktop virtualization helps to increase the student-to-computer ratio, which is very significant in a twenty-first-century learning environment. From the standpoint of an owner of an Internet café, virtualization increases users' access to computers/Internet without purchasing additional computers.

While desktop virtualization has advantages, it must be known that some proprietary software still has issues as far as licensing is concerned. If the platform is open source (Linux), then licensing will not be an issue by virtue of the open source model.

As I mentioned earlier, desktop virtualization used at an academic institution like B. W. Harris Episcopal High School increases the student-to-computer ratio. At a time when the school, through the support of its alumni association, was being transformed into a twenty-first-century educational environment driven by technology, access to more computers with Internet capability was not only imperative but was critical. Hence, the installation of virtualized computers was an economical means of providing this increased access to the Internet. The installation at B. W. Harris School included one computer system, serving as the host, connecting to three other monitors attached to three terminals serving as clients, thereby increasing the total number of workstations to four. The desktop virtualization kit used was purchased from nComputing.com. In the future, it is hoped that another setup giving thirty students access to a desktop by simply using *one* computer as the host will be installed. This will be done by using another virtualization product from nComputing.com.

Generally speaking, setting up a fifteen-workstation computing environment using desktop virtualization will only require one high-performance computer, sixteen monitors, sixteen keyboards, sixteen mice, sixteen speakers (for sound), fifteen virtualization terminals (from www.ncomputing.com), and sixteen network cables (Cat 5, 5e, or 6), and a 24-port switch. Since the host computer will be performing server functions, we can only count the fifteen workstations as those available to users.

It is important to note that this setup would be ideal for libraries, computer labs, Internet cafés, and schools. It is not recommended that these terminals be used in government offices, businesses, or any organization where security and reliability are paramount unless they are thoroughly tested.

The Future

As Liberia continues on the path of development, many large companies will begin to invest there. These companies will run data centers that may not consume as much energy as traditional data centers because of advancements in technology. Most companies directly investing in Liberia might gravitate toward the cloud-computing paradigm for their computing needs because this would seem a safer approach to computing in an African nation that has a history of instability. Moreover, the need to cut costs in capital investment will force IT departments in Liberia to adopt virtualization. Also, the use of energy-efficient devices will be paramount in IT department policies and operations.

A "Green Initiative"

Maintaining a Green IT environment requires a legal and regulatory framework that would guide the process. Without policies it may be difficult to get all stakeholders on the path that leads to such an environment. All firms or organizations in Liberia, including the government, will need to engage in a massive campaign to "go green." Internet cafés need to run greener environments. Schools and organizations that receive shipment of technology items from their alumni association or mother or sister organizations must ensure that they communicate the need for the donation of greener technology equipment. The National Port Authority and the Bureau of Customs must also ensure that there are policies in place that will reject "nongreen" equipment. Once this becomes a national initiative, IT departments in Liberia stand a better chance of being green.

Liberia needs a "Green Initiative," which is an effort to advocate, educate, and implement the greening of Liberian IT departments in order to save energy and the environment. The initiative should be in the form of an agency or commission that has a leadership with a vision for a "Green Liberia." This entity will ensure that the Liberian IT sector follows the best practices of a green environment. These practices should include the hiring of a Green IT specialist to lead the effort in counties and organizations, requiring IT departments to consolidate and virtualize their data centers, encourage the installation of energy-efficient devices, and so on. This sounds like a task for the Environmental Protection Agency of Liberia.

Finally, with a green environment, we can eliminate the need for an infrastructure that requires increased power and build a more energy-efficient one. Doing so will decrease our dependence on fossil-fuel energy. We can also save a lot of money that we would have otherwise paid for fuel to run diesel engines. IT professionals must help champion the "Green Initiative" throughout the country. IT professionals are the decision makers as well as those who design, configure, install, and maintain these data centers or those IT environments that require their expertise. They are the ones with the most power to ensure the success of a "Green Initiative" in a digital Liberia to protect the Liberian environment. That, to me, is a social responsibility!

CHAPTER 29

WEB 2.0: ITS IMPACT ON A DIGITAL LIBERIA

The term Web 2.0 was coined by Dale Dougherty in 2004. Later, Tim O'Reilly, of computer software manual publisher O'Reilly and Associates, popularized the term in several of his articles. O'Reilly has been and continues to be a very influential figure in the ICT world.

Web 2.0 fulfills Tim Berners-Lee's original vision of the World Wide Web. Berners-Lee is considered the father of the World Wide Web. Had it not been for Berners-Lee, I doubt that many of the developments we have on the Internet today would have occurred.

Web 2.0 is now a popular term used for the way we now use the Internet and involves the use of advanced Internet technologies and applications that enhance interaction, collaboration, participation, and creativity. It has changed the Web from a medium to a platform where online communities exist in a virtual world. It has further broken geographical boundaries and made computing over the Internet much more engaging. Most of all, it provides tools that certainly discard the need to learn computer programming to create a product, which gives everyone unprecedented capacities.

Web 2.0 services are built on AJAX, which consists of an amalgam of Web-page coding standards such as JavaScript and XML, allowing direct interaction with Web pages. It takes the user from the grip of a personal computer into a virtual community independent of the limitations of the personal computer. Web 2.0 allows the user to use the Internet as a personal computer. It is inherently social and promotes interaction and participation.

As I mentioned earlier, Web 2.0 changed the way we interact with the Web. The original Web (Web 1.0) consisting of static Web pages seemed to be more of a medium, while Web 2.0 brings forth a dynamic platform that encompasses a new paradigm. The term Web 2.0 distinguishes itself from previous generations of Web applications in many ways but specifically because it is interactive and collaborative.

Web 2.0 has brought several new resources that have pervasively impacted the global spectrum. I do not spend too much time on all the resources provided by Web 2.0 in this book, especially in this chapter, except those that have had a sudden and an extensive impact on society. I list a few of the resources below:

- Social networks
- Blogs
- Wikis
- RSS-generated syndication
- Mashups
- Video sharing
- Bookmarking

In this chapter I discuss Web 2.0 and the impact it has and will have on Liberia and not on the details of its (Web 2.0) function. In Liberia and other developing countries, the impact of Web 2.0 is already being felt. The use of social networks like Facebook, Hi-5, MySpace, YouTube, etc., is common. Yet, the general public is not even fully aware of what Web 2.0 is really about or the benefits that this new "phenomenon" provides for a country that is behind in terms of technology.

Web 2.0 has and will continue to change the way many things are done in the Liberian society. The way education will be delivered will change, politics or political campaigning will change, businesses will change, national security will change, and the health sector will change. The question is, will Liberians be ready to leverage this new phenomenon for the betterment of their country?

Web 2.0 As A Tool In Liberian Educational System

In chapter 7, where I discussed distance learning as a way to improve education in Liberia; I mentioned Moodle, Blackboard, and WebCT, which are all learning-management systems currently being used by educational institutions around the world. While these software systems are good for distance learning, using the Web 2.0 approach takes learning to another level. It makes distance learning interactive, collaborative, exciting, and global. The use of blogs, wikis, social bookmarks, chat, video blogs, and conferencing brings excitement and participation to learning, all of which are characteristics of twenty-first-century educational paradigms. Also, professional development can be enhanced by using Web 2.0. Teachers can form "teachers' social networks" or maintain their own blogs that allow teachers around the world to interact with them to enhance their instructional approaches and broaden their horizons. This medium also creates information exchange, which ultimately leads to a knowledgeable society.

Since Web 2.0 allows active participation, it will force both the instructor and the student to think critically, which needs to be incorporated into the Liberian school system. Web 2.0 for education will work perfectly in Liberian schools and will change the dynamics of education in Liberia.

Another area in which Liberian academia can benefit from Web 2.0 technologies is the area of research at the university level. Included in Web 2.0 technologies are applications and services that allow the individual to create, store, and publish his/her work, allowing it to be viewed, critiqued, or used by others. This is a good opportunity for university students and researchers and professors to publish and share their works with the global community. Web 2.0 has a global "reach" and encourages participation in the delivery of information. This makes garnering information in a timely manner very easy; not to mention increasing the diversity of contribution to research work.

Web 2.0 will enhance education in Liberia by doing the following:

- Incorporating twenty-first-century techniques into the classroom
- Empowering students by providing multimedia resources
- Improving differentiated learning
- Encouraging the use of digital media to enhance education

Institutions of learning in Liberia need to begin investing in and adopting new approaches using Web 2.0 technologies to change their methods of delivering education. Liberian instructors should start thinking about running entire classes digitally using social networks to engage students and create a participatory environment where there can be an exchange of ideas. Educational administrators must begin to integrate Web 2.0 technologies into their strategic plans and offer the support that teachers need to provide good education to Liberia's future leaders.

Web 2.0 provides many applications that can be integrated into the educational environment. But most importantly, Web 2.0's API (applications programming interface) and data are open source, which makes it easy for users to create and integrate applications tailored to suit the Liberian educational system, business, private and public sectors. Our ability to leverage Web 2.0 to improve our educational system requires some work and openness to a new vision.

Web 2.0 In The Liberian Government

In the government and public sectors, obviously, Web 2.0 will enhance transparency and democracy. But most of all it will encourage national participation in government affairs. In politics, especially in elections, Web 2.0 will bring a revolution in Liberia as it has done all over the world. You may remember that on February 10, 2007, when Barack Obama first launched his political campaign in Springfield, IL, many

Americans believed that his campaign would be yet another failed attempt by an African American to seek the most powerful position in the world. But Obama defied all odds and brought a paradigm shift in modern political campaigning. Obama's shift from the status quo to the use of ICTs is now being emulated in presidential elections in other countries.

In South Africa, President Jacob Zuma's resounding electoral victory was attributed partly to the effective use of technology. At the time, President-elect Zuma was said to have used the Nokia N95 cell phone as well as a laptop connected to a 3G network to campaign whenever he traveled throughout the country (IT News Africa 2009). This is yet another example of how ICT has influenced electoral politics.

Candidate Barack Obama's use of technology weighed heavily on a Web 2.0 paradigm. Facebook and YouTube were used very effectively and were able to garner more votes for Obama, especially from young voters. It was obvious that Obama was bringing a change to electoral politics and had targeted a large segment of society that is often overlooked, the young voters, college students, and all of those who have become regular users of social networks. The result of this use of Web 2.0 technologies as we all know was the election of the first African-American president of the United States. Today, as president, Obama still relies on social networks to reach out to the American people and the world; something that is being done very effectively.

Candidates entering the electoral process will have the opportunity to gain competitive advantage by effectively using Web 2.0 technologies to advance their political philosophies and even raise funds. The issue now is whether they are willing to evoke such an approach in their political campaigns.

In 2011, Liberia is expected to hold its general elections, which are expected to be very competitive considering the plethora of competent and potential candidates that are aspiring to run. Already aspiring candidates who are young and are consumers of social networks are beginning to use these networks to create awareness for their imminent campaign endeavors. As political campaigns shift from the traditional approach toward the modern approach of bringing information to the people via social networks, candidates seeking electoral office in 2011 should be willing to effectively leverage the power of Web 2.0

technologies to propound their political philosophies. This approach will impact the Liberian electoral process, because Liberia's young voting population have become heavy users of these social networks and are enormously unfamiliar with the old political atmosphere in Liberia.

Web 2.0 For Fundraising Purposes

Web 2.0 technologies can be very useful in raising funds, since social networks can be a cheap and very effective tool in fundraising initiatives. During the presidential elections, then-presidential candidate Barack Obama raised half a billion dollars online using e-mails, texts, and social networks (Vargas 2008). Social networks were leveraged to fuel this unprecedented level of donation. This situation can be applied to the ensuing Liberia elections or any form of fundraising initiative; Liberians all over the world would be able to contribute to a cause and make monetary donations through an online payment system that is aligned with a banking institution. This is an unprecedented initiative in Liberia but has the potential to produce lucrative results.

Other Areas That Will Be Impacted By Web 2.0

I spent a lot of time discussing the impact of Web 2.0 in education and politics because these two areas have been affected greatly and globally by Web 2.0. But Web 2.0 can have the same impact on other areas as it has on education and politics globally. For example, in the security sector, security personnel will be able to develop social networks that link to their collaboration mechanism to garner information or use the art of deception to track criminals.

In the health sector, the Ministry of Health could create a health social network that would allow the participation of individuals locally to contribute solutions globally to problems that may require local expertise. With global participation, collaboration could be enhanced to achieve results.

The print media, which has already begun utilizing the Web, will have to be more innovative in the use of Web 2.0 and leverage it to replace the print section of its business. Print media has felt the impact of the Internet in terms of declining newspaper sales because most of their customers have gravitated toward the Internet to read online news, which is often more up-to-date.

Web 2.0 will empower Liberians in many ways. The use of blogs will allow Liberians to publish their own content without paying exorbitant prices, as was done in the past. The availability of free and user-friendly blogging media such as the one offered by WordPress.com, levels the playing field for every Liberian who needs to get a message out to the public. Blogging has brought a revolution to publishing content and will greatly benefit Liberia.

Web 2.0 will enhance participation greatly in Liberia. It will change people from consumers to "pro-sumers" and will bring about an explosion of innovation in every aspect of the Liberian society. This explosion of innovation will be a result of available knowledge and collaboration; two things that the country has lacked over the years.

Before we move on to Web 3.0 or Web 4.0, there are many things that their predecessor Web 2.0 will do that will change the way all Africans, not just Liberians, use the Internet.

Web 2.0 will force transparency because it will eliminate the authoritative grip a select few held and give power to all. In education, the scenario will gravitate from knowledge emanating from the expert to knowledge shared in a medium where the role of expert changes to facilitator. In a typical learning environment, which previously was passive and boring, learners will gain a new passion for learning because of its participatory nature. Schools will not be a place where students just obtain a formal education but a place where every student will garner lifelong learning. Overall, Web 2.0's impact on Liberia will be revolutionary.

Chapter 30

Cloud Computing—A New Utility And The Genesis Of Ubiquitous Computing In Liberia

There has been much discourse and literature about cloud computing as the computing paradigm of the future. Several firms, industries, governments, and academia have begun to adopt this new computing paradigm. What is more interesting is that cloud computing is being considered the ultimate solution to ubiquitous computing in developing countries, thereby bridging the proverbial digital divide (Wilkins 2009). Furthermore, in the summer of 2008 International Business Machines (IBM) announced the implementation of cloud computing in South Africa (Innovation Africa 2010). That initiative signaled the genesis of ubiquitous computing and a potential enhancement of ICT penetration on the continent. But despite this optimistic endeavor several factors have been identified in African countries that may restrict or strangulate a cloud-computing implementation. The lack of ICT capacity, lack of infrastructure, lack of political will, and heavy taxation on ICT products are a few of the most frequently mentioned factors in discourses relating to ICT penetration in developing countries, including Liberia (Wilkins *Liberia Daily Observer* 2010e).

So what is cloud computing? How can it be applied to or how can it benefit Liberia? This chapter discusses cloud computing as a new form of utility that can be provided to Liberians by both the government and the private sector and how it (cloud computing) can enhance ICT penetration through ubiquitous computing in Liberia.

Basically, cloud computing is computing over the Internet using Web services and applications. It involves a lot of scalable and virtualized resources, and everything is pretty much provided as a service. Cloud computing is user-centric, task-centric, powerful, accessible, intelligent, and programmable (Miller 2009). It is the superstructure that leverages omnipresence, always-on connectivity, spawning a new computing paradigm that involves software as a service (SaaS), infrastructure as a service (IaaS), and my favorite, platform as a service (PaaS). Software as a Service is when a provider licenses software to a customer for a fee. The customer only pays for the software that was licensed and nothing else. Infrastructure as a Service follows the same approach as SaaS although, in IaaS an entire infrastructure is licensed for a fee. An infrastructure includes virtual servers, network switches, connection, etc. Platform as a Service or PaaS involves the licensing of a platform to customers. Platform licensed from providers are used to develop and deliver software.

"The shift to cloud computing is fueled by the dramatic growth in business collaboration, connected devices, real-time data streams, and Web 2.0 applications such as streaming media and entertainment, social networking and mobile commerce" (Innovation Africa 2010).

Cloud computing eliminates the need for a customer/user to spend resources on assets such as software and hardware. For example, if you open a small office, you would need capital to purchase computers, software licenses, etc. This can cost a lot and have a huge impact on your capital investment. With cloud computing, you would not need to purchase software or a lot of hardware (servers, etc.) to run your own ICT infrastructure. While you will need some low-cost hardware (computers or terminals), your entire ICT infrastructure will be provided by "the cloud" (which is the data center of a provider), and you will be billed for usage, storage, security, and so on. This means that your ICT needs will be provided as a service. You see, cloud computing applies a model known as utility computing, which parallels the model used by companies that provide the utilities that are found in homes, such as electricity (Liberia Electricity Corporation), water (Liberia Water and Sewer Corporation), Internet services (Comium, CellCom), and cable

(DSTV). In cloud computing, customers are billed on a subscription basis and only pay for what they use.

Cloud computing has become very popular in all parts of the world: academia, business, government, etc. In the business world, cloud computing has been adopted by several large firms and has been providing services to a large base of customers. IBM, Yahoo! Google, Amazon, Microsoft, and Dell, have already engaged in this computing paradigm (Schonfeld 2009).

Cloud computing comes with a lot of promises, but what seems to be very encouraging are its promises of universal access, 24/7 reliability, ubiquitous computing, and above all, the reduction in computing cost. Cost is a major issue for a country like Liberia when considering infrastructure. The possibility of running computing services without investing heavily in an IT infrastructure indicates that countries like Liberia are getting closer to bridging the proverbial digital divide and getting ready to engage in the digital economy.

Since it is a utility that can be provided to customers/users, Cloud computing can be implemented in Liberia by the government, the business community, and individuals. In the next paragraph, I will discuss how the government can provide cloud services as a public service as well as a commercial service. I will also discuss how businesses can go about providing cloud services to customers.

Government Providing Cloud Computing As A Utility

To provide public services for government-owned institutions such as the support for an e-government infrastructure, the government of Liberia will need to invest in a modern state-of-the-art data center. This data center should host the national cloud-computing infrastructure. It should include several servers, applications, networking, storage, and other devices that operate in a virtual environment. It will support the Government of Liberia's "e-government" services as well as institutions of learning, public and autonomous agencies, the Liberia School System, etc., in Monrovia, as well as government institutions in all fifteen counties of Liberia. LIBTELCO, which is the country's national operator, would be a likely candidate to provide these services.

On the other hand, the Government of Liberia could provide commercial computing services on a subscription basis to the public through LIBTELCO. But I think things will work much better if an autonomous agency is established to provide dedicated cloud services to customers. This autonomous agency could be named the Liberian Cloud-Computing Corporation (LCCC) or any other name that the government chooses. The company will provide cloud services to customers in the same way other utility companies like the Liberian Electricity Corporation does. LCCC will provide the necessary equipment and materials needed to facilitate computing. This equipment and materials include "cloud computers," networking equipment, and so on. Technical support, training, and other services such as Web development, e-commerce, etc., can be provided to generate additional revenue.

This new agency will require qualified professionals to run data centers. This will require an investment in human capacity, which can be garnered from the Diaspora. There are several ICT professionals in the Diaspora. These professionals make up Liberia's ICT brain gain, which can be used to ameliorate Liberia's ICT sector. Instead of investing in expatriates, the government should provide the incentives that can lure Liberians from the Diaspora back to Liberia to help rebuild the country.

Business Sector

The prospect of broadband coming to Liberia through the ACE fiber-optic cable system opens opportunities that can be explored by everyone in the business sector. Businesses and individuals who see opportunities and potential in cloud computing can open data centers and provide services to subscribers. Companies that are most likely to engage in this sort of business might be existing Internet service providers (ISPs), mobile operators, cable companies, and other businesses that already have the infrastructure and capacity. These companies will merge services and market them in packages. They will combine cloud-computing services with cable, Internet, and phone into one package and refer to it as "Quadplay," or some other name. This will allow subscribers to pay a single amount for an entire package that includes voice, video, data, and computing. In this case, "computing" will include the use of applications, hardware, and services involved in the "cloud." These devices (computers, laptops, mobile devices) would be devices manufactured specifically for cloud computing or existing devices.

A typical example of a computer that could be considered a "cloud computer" is the atom-powered mini-computer known as the Netbook. Netbooks are basically designed for the Internet; to learn, communicate, and view information. Several companies have begun making the Netbooks—Acer, Hewlett-Packard, One Laptop per Child (OLPC), Dell—and as cloud computing continues to develop and mature, more "cloud computers" will emerge. Low-powered Netbooks and terminals that use fewer resources intensive for processing power will suffice in regions that have lower ICT penetration like Liberia. Netbooks run for up to nine hours (Biersdorfer 2009). Intel is also making low-power processors for mini-desktops called Nettops that would also be ideal for cloud computing (Gonsalves 2010).

Ubiquitous Computing

Ostensibly, cloud computing levels the playing field for individuals, small companies, and countries to compete with bigger ones in terms of innovation and creativity. It allows an ordinary computer programmer situated in a remote area like Tchien, Grand Gedeh County, who does not have a programming platform to use "Platform as a Service (PaaS) to compete with large companies like Google, Sun Microsystems, IBM, or Microsoft. It will also allow a student in Maryland County to be able to access online books or resources even though her county does not have an ICT infrastructure. The county health officer in Weasua or other rural areas will have direct access to the Ministry of Health's database as well as the database of the John F. Kennedy Medical Center.

Liberians will be able to use Netbooks or devices that have the same characteristics as Netbooks to connect wirelessly through an access point to resources available in the "cloud." For example, a student living in the rural area can use a Netbook to access MIT OpenCourseWare and use its available resources as a supplement to resources provided by Liberian schools. This will allow students to garner the type of intellectual exposure needed to work in the digital economy and compete with their contemporaries from developed countries. IBM and Canonical (developers of Ubuntu Linux) have begun to introduce "new flexible personal computing software package for Netbooks and other thin client devices to help businesses in Africa bridge the digital divide" (Chairman King n.d.).

What Does Liberia Need To Implement Cloud Computing

Liberia needs five basic components to implement cloud computing: electricity, a broadband connection, a cloud infrastructure, human capacity, and policies. Electricity is being provided by the Government of Liberia and being sold as a utility. The country's provider of electricity, the Liberia Electricity Corporation, continues to make significant progress with the help of the Chinese government. For broadband connection, the ACE fiber-optic submarine cable is expected to be ready for service (RFS) by the second quarter of 2012 and will be managed by LIBTELCO, LTA, and other mobile operators. As far as the infrastructure, as I mentioned earlier, LIBTELCO or preferably an autonomous agency should provide the infrastructure. Th e Diaspora will be relied on for human capacity, while the Ministry of Post and Telecommunications or the Liberia Telecommunications Authority will handle the policy and regulatory aspects respectively.

A Few Challenges

Some challenges that need to be addressed include but are not limited to capacity building, pricing, and a high poverty rate, which may limit subscription to the "cloud" and strangulate its market prospects.

Security is another major issue since Liberia is an African country emerging from a prolonged period of war with the potential for political instability. Also, corruption and the lack of qualified professionals to run a sustainable cloud infrastructure could impact its security if Liberia's "brain gain" (Diaspora) is not fully utilized or neglected.

For cloud computing to be accepted, a better understanding of the aspects of security and privacy regarding a users' interaction must be provided by those offering cloud services. As soon as this issue is addressed, more users will gravitate toward cloud computing. Despite these concerns, Liberia shows great potential in terms of running a successful and sustainable "cloud" infrastructure, although a decision to implement cloud computing in Liberia must be based on thorough research and a "pilot" implementation to ensure success and growth.

Finally, in more than six years of studying ICT for development, two technologies strike me the most when it comes to bridging the digital divide: mobile technology and cloud computing. While I believe that the former has a greater penetration, I also believe that the latter will be responsible for a greater social and economic change in Liberia and other less-privileged countries.

CHAPTER 31

A MOBILE NATION

Information and communications technology have over the years experienced several revolutions: the mainframe revolution that occurred between the '50s and the '70s; the minicomputer revolution of the '70s and the '80s; the personal computer revolution of the '80s and the '90s; and the networking revolution of the '90s and the 2000s that led to the Internet, which has connected the world. Now, in the twenty-first century, we are experiencing yet another revolution—the mobile revolution. Each of these revolutions has had a significant impact on every aspect of mankind: economics, geography, health, education, politics, and so on. The most recent revolution, the mobile revolution, has had an enormous impact on developing countries by literally bridging the digital divide that has for so long separated the "Dark Continent" from the other parts of the world. Liberia is one of the many countries where the impact of the mobile revolution has been felt.

In this chapter, I discuss the general impact that the mobile revolution has had on Liberia; placing emphasis on the use of mobile phones and how pervasive it has become in the rural sectors. I then discuss its impact on the Liberian economy and end with its impact on Liberia's future.

Since the end of Liberia's fourteen-year civil war, the country has experienced many changes and relatively new developments in the area of telecommunications: the advent of mobile phones and operators, access to the Internet via satellite, the establishment of the Liberia

Telecommunications Authority (LTA), the revamping of the Liberia Telecommunications Corporation (known now as LIBTELCO) and its accomplishments, the drafting of an information and communications technology (ICT) policy document, the opening of new information and communications technology schools/institutions, the introduction/implementation of the ATM machine, and many more new developments that have enhanced ICT penetration in the country. These are remarkable achievements for a country emerging from a prolonged and devastating civil war. Of all these new developments mentioned above, mobile technologies have had the most revolutionary impact on Liberia, especially the cellular phone.

Mobile technologies have advanced in Liberia and will continue to advance beyond our expectations. They have and will continue to penetrate the country and will continue to be a vehicle that drives the economy. Mobile technologies will not only drive the economy or the telecommunications sector but will make Liberians more connected to the global community as well as other technology resources. This technology will turn a country that was once a nation that relied on obsolete donated computers to one that relies on portable computing and communications devices. Those portable computing and communications devices will have access to resources and perform tasks that their predecessors (desktop computers, analog phones) could not perform. Computers will work like phones, and phones will work like computers. Liberians will turn toward those portable and mobile devices. The smart phone will become pervasive in Liberia and will be used for e-mail, chat, Internet browsing, and other functions other than for making phone calls. Cellular phones will change from a single purpose luxury device to a multipurpose device of necessity.

The mobile impact will be felt greatly in the healthcare sector because of the introduction of new technologies, especially those that provide video access. It will enhance telemedicine, which was discussed earlier. Liberians will live longer because with wireless and mobile devices; patients in rural areas will have access to doctors in the modern cities. The use of video technologies via mobile devices will give doctors in remote areas access to their patients in places where they (doctors) may not be able to reside. Body area networks and other wireless devices will improve health-monitoring services and allow doctors to test and diagnose patients from remote locations.

Smart phones—the Blackberry, Palm, iPhone, and so on—are considered mobile computers since they are built with processing power and applications that parallel the personal computer.

With the maturation of the Automated Teller Machine (ATM) in Liberia, businesses will migrate toward "e-commerce and m-commerce" to gain competitive advantage. With m-commerce, Liberians will be able to purchase consumer goods, monitor their bank accounts, and do most, if not all, of their business transactions over the Internet using their cellular phones or any Internet-capable mobile device. The Central Bank of Liberia's plans to open a stock market will be fueled by mobile technologies (*n.a.* 2009). Participating investors will need access to the stock market and access to the market, and especially in the rural sectors this will be facilitated by mobile devices. The financial sector will leverage the Internet through mobile technologies to add value and gain competitive advantage. The proliferation of banks in Liberia will further escalate the need for a flexible information technology environment that can adapt seamlessly to inevitable changes brought by mobile technologies.

Engineers in Liberia will not have to carry large plans and other bulky resources used during construction. Portable devices will be powerful enough to carry CAD (computer-aided drafting) programs that will have plans and other diagrams that are used in the construction industry. Wireless portable devices with access to data centers or cloud computing centers will be used in place of traditional computing devices such as the computer.

The same will work for farmers! Cell phones will play a major role in the agriculture sector. Unfortunately, in the agriculture sector majority of Liberians do not have formal education; hence, using their portable devices as a computer would be challenging. Therefore, superintendents, cooperatives, the Liberia Produce and Marketing Corporation, and the Ministry of Agriculture will have to implement a means of communications that will allow farmers to get their produce to the market. Automated dialers should be used to send "robocalls" to farmers (in their respective dialects, of course), informing them of where they can call to have their produce picked up or where they can deliver their produce. This would be m-agriculture, or mobile agriculture—the use of a mobile device to aid in the area of agriculture.

In politics, mobile technologies will bring change to the electoral process. Traditional electioneering will be supported or overwhelmed by electronic electioneering, and Liberians will gravitate toward m-electioneering because of the successful penetration of mobile technologies.

Mobile phones will make communications between candidates and their constituents convenient and seamless. Travels to constituencies will be reduced but certainly not canceled. This is because the divide between those in the cities and those in the rural areas has been bridged by mobile technologies. Every candidate will have equal opportunity to reach the voters in the rural areas. Candidates will now be required to actively campaign for votes, unlike in previous years, when there were limited or no means of communication and voters in the rural sectors were confined to hearing the views of only those candidates who were able to reach them. Today, a person in the rural sector does not have to base his or her decision on the fact that he or she received a bag of rice from a candidate; instead, voting decisions will be based on the candidates' views and philosophies and what he/she can do for the voter's county and the entire country. Amazingly, voting decisions might most likely be based on what a voter hears from a relative living in the Diaspora, or other parts of Liberia.

I gave an example of this in one of my articles sometime at the beginning of 2010. In that example, I mentioned my Aunt Rachel, who resides in Pleebo, Maryland County. I gave a hypothetical situation in which a friend of mine was a candidate aiming for an elected office in Maryland. Knowing that my friend needs the votes of the elders and any other group of voters that my aunt has influence over, I would call Aunt Rachel and convince her that voting for my colleague would be the right thing to do and would be in her best interest. Knowing my Aunt Rachel and how she feels about me, she will not only give my colleague her vote, but she will also convince her friends to do the same. Now, remember, I lived in the United States and Aunt Rachel lives in Maryland, Liberia. Moreover, I had not seen Aunt Rachel in years but only got in contact with her when she received her mobile phone. So this is an example of the impact mobile phones will have on the elections.

Mobile technologies will also impact our democracy especially press freedom, human rights, and so on, which faced several violations in the past. For example, if a news crew comes to a country to cover a story and gets arrested, the world will know in a few seconds. This is because body area networks, or BANs, have made it possible for individuals to attach hidden devices/cameras on their bodies, which can transmit to other devices or networks at distant locations in real-time. These devices can come in the form of a pair of sunglasses or a pen, a fake tooth, or a belt buckle. With these wearable devices on, an incident of brutality perpetrated against a member of the press crew can be broadcast in real-time. Before the perpetrator realizes it, he/she would have already been televised on CNN, BBC, MSNBC and/or the Internet (You Tube, Facebook, MySpace, etc).

In the security sector, mobile devices will benefit members of the joint security consortium assigned to border areas. These devices, as I mentioned earlier, will have access to a system like CENTINOL, allowing collaboration and information sharing. Portable and mobile devices will be used to fingerprint criminals or suspects, and information can be e-mailed via a handheld device to a central database for processing and immediate response. Intelligence officers will use BAN devices to perform intelligence operations. These devices will connect to a central database for information processing and sharing.

Educators have yet to find innovative ways to use mobile phones in classrooms even though they (mobile phones) have changed the culture of our society. In Liberia, almost every student has a phone, which, from what I gathered, they answer during class. Schools have policies regarding the use of cell phones in classrooms. Some, if not most of these policies do not favor students; hence students rebel in their own ways. This rebellion can also impact student achievement in schools, so teachers must search for new and exciting ways to teach and be able to integrate cell phones in their lesson plans. Schools need to make policies flexible in order to be able to inject cell phones into the classroom although with some rules guiding the use of them; basically for educational uses *only*!

One way teachers could influence the use of cell phones in the classroom is to collect a list of student numbers and make sure class work, homework, or any academic assignment is e-mailed or sent by text to students. Encourage utilizing their cell to discuss the assignment that was previously sent. This will create excitement and improve student learning.

Another use of mobile technology is through the grade-book system where grades are pulled from the school's system and sent to students' and parents' phones. Basically, this would be a home-to-school communications technology solution. This will engage parents in their children's learning process. There are many uses for mobile devices in school. Since cell phones have penetrated every segment of the Liberian society, this should be leveraged for education to enhance Liberia's twenty-first-century educational initiatives.

In early 2010 when I was in Liberia, smart phones were gaining popularity. And it seemed they were going to be the phone of the future. I noticed one thing at the time though—a lot of people who carried smart phones only carried them because they looked "cool" or "classy" or simply because they distinguished them from those of us who carried the USD $20 phones. But they barely use their smart phones for anything other than making calls—not for Internet browsing, chat, or even Facebook, the newest phenomenon in Liberia. I even saw a T-Mobile smart phone from the United States that had a global positioning system (GPS). I decided to test the GPS feature of the phone by entering my address to get directions, but that did not work.

Cellular phones will further penetrate the entire Liberian environment and will be the medium through which Liberians will connect, communicate, collaborate, and compute—just as they are doing now. They will grow to be multipurpose devices. Liberians will be able to use their cellular phones as radios and televisions.

Most of the cellular phones that are used today are built using the open source Linux operating system. This makes the development of more communications and/or collaboration software customized for Liberia possible. Palm, Microsoft, and Symbian are the dominant producers of smart phones. We should soon see them compete for business in Africa.

The global telecommunications industry has faced several challenges in the past years, especially at the beginning of the twenty-first century, but it has been very resilient. This resilience emanates from the result of its research and development efforts, which has brought great changes to the world. Considering the progress that has been made in mobile technologies globally and in Liberia, I can confidently say that their (mobile technologies) impact on Liberia is still yet to be felt. Mobile technologies will penetrate Liberia faster than information technologies. But both mobile and information technologies will combine to make what will be a digital Liberia.

Epilogue

After several trips to Liberia and endless nights of contemplation, the concept of this book was conceived. What I had gathered from my trips to Liberia was hope; hope that Liberia would soon transform to a better and more modern nation. And the more hopeful I became, the more I garnered a renewed motivation and enthusiasm to contribute toward Liberia's transformation. But despite my passion and resolve, I am a Liberian in the Diaspora guilty of not being able to leave his family in the United States, especially my ailing mother.

Moreover, the thought of permanently relocating to Liberia without gainful employment capable of covering my personal responsibilities was even more frightening. So I made two futile attempts to secure jobs in Liberia in hopes that my extensive ICT and business background would give me a competitive edge over other applicants. I had forgotten that in these days, getting a job remotely or even locally is not always about what you know but oftentimes who you know.

My unsurpassed intransigence would not allow me to shift to the despair side of the pendulum but rather toward the side of optimism and perseverance. And this further motivated me to somehow continue my journey. I was resolved to not give up on Liberia. I felt and still feel that Liberians in the Diaspora are Liberia's "brain gain "and the ones with the resources to start the revolution. To give up simply because my first attempts to secure employment or engage in other private and personal endeavors were unsuccessful would essentially mean giving up on Liberia. And then, would I be any different than other Liberians who are yet to see the big picture? That would make me as guilty as they. This is not a matter of opinion; it is a matter of fact.

I have come to the conclusion that the time has come to contribute to Liberia in any way I can. After all, it is my motherland. So after months of meandering back and forth over what would be a major contribution, I finally decided to contribute by writing this book! I am confident that the ideas and solutions proffered in this book will help in transforming Liberia into a knowledge-based society.

Several months have passed since I began writing this book, and many things have transpired. There have been conversations/discussions about ICT capacity building, the construction of the International Telecommunications Union (ITU) funded Training Center of Excellence in Voinjama, Lofa County, more computer schools have opened, and there has been improvement in the energy sector, with the Liberia Electricity Corporation providing power partially to the city of Monrovia. Liberia's mobile teledensity has also increased and there's a potential for progress. A new University of Liberia campus was constructed by the Chinese Government at Fendell, while several others are being constructed around the country. A new modern hospital was also constructed by the Chineses Government in Tappita, Nimba County.

What is arguably of great importance is that two weeks after I wrote an article relating to the ACE submarine fiber-optic project in the Liberian Daily Observer, the Ministry of Post and Telecommunications published a release on the very subject. The release was a "Policy Framework" for submarine fiber-optic cables in Liberia (Ministry of Post and Telecommunications 2010). The focus of the press release was on the ACE project. This policy framework it seems, may finally be the initiative that will lead to the beginning of the digital transformation in Liberia which I referenced several times throughout this book.

Following that press release mentioned above, on June 5, 2010, Liberia joined twenty-five other countries in Paris, France, in signing the ACE Construction and Maintenance Agreement contract with France Telecom. The signing of that agreement will give Liberia access to the ACE submarine fiber-optic cable system and the installation of a landing station in Liberia. "The station landing in Liberia will be owned by a Special Purpose Vehicle called the Cable Consortium of Liberia

(CCL), a newly incorporated consortium comprising of LIBTELCO, LONE STAR, and CELLCOM"(TLCAfrica.com 2009). This was good news, and as always, I commended the Liberian authorities responsible for this development. I also reminded them that joining the ACE consortium was just the beginning and that there was more work to be done to achieve the goal of bringing broadband to Liberia. With the Government of Liberia receiving a waiver on its US $4.9 billion debt (Afrique Avenir 2010), there is hope that the country's ICT sector and all other sectors will soon improve.

There are many things that I mentioned in this book that will soon begin to occur. With the ACE project in sight, I look forward to seeing the formation of a National ICT agency/division/department that will centralize ICT expertise and be responsible for integrating ICT in the country. This entity will also bring an explosion of innovation that would make Liberia a viable competitor in the digital community.

The Ministry of Health should either adapt or implement the Health Information Technology Hub (HITCH) that I mentioned in this book in addition to the myriad of proposals that I listed. If it chooses not to implement my suggestions/recommendations on using ICT in providing healthcare, then I hope it implements something that closely resembles them. A tele-health program will benefit the Liberian healthcare industry and will improve the health sector enormously.

Other important initiatives would include the Ministry of Education's engagement in a massive and robust effort to build capacity nationally, and the development of an educational technology plan that revolutionizes Liberia's educational system. The education sector needs to be revolutionized and not just reformed. I have discussed several approaches in this book, which I hope will be used as a guide. I cannot stress the importance of implementing changes in this sector, because it is critical to Liberia's future. Below is a list of additional ideas that would be beneficial to Liberia.

1. Develop an Intranet that resembles LEARN (Liberian Education and Resource Network). LEARN (www.LEARNLIB.Org) is an open source project that I started to help Liberian students find ICT resources as well as local and international scholarships. But what I am suggesting in this book is a medium that provides repository of

educational resources for students and teachers as well as access to electronic resources including online databases.

2. Create technology education centers in local communities to allow students who attend schools that cannot afford to run their own computer labs to have access to computer and the Internet.

3. Create a mobile computer center that travels to schools that cannot afford a computer lab, especially in the rural areas. Netbooks or other computers require little resources and energy would suffice for this initiative. This is similar Uganda's UConnect initiative.

4. Include access to television in schools so that they can have other options for learning. This means schools must be renovated or built with outlets for Internet access, granting them the capacity to run data, video, and voice access.

I discussed e-government and how it will impact Liberia when fully implemented. But I would like to recommend that counties get their own sub-domain within the government's e-government infrastructure. The format should be www.GOL.County.Lr, where "GOL" stands for Government of Liberia, "county" will be the name of the county and ".Lr" stands for Liberia. This will allow each county to run its own Web site and services although guided and monitored by a central government ICT agency. This will give counties independence, enable them (counties) to provide residents in Liberia and the Diaspora with pertinent information, and provide the global community a one-stop shop for county-specific information regarding investment opportunities, landscape, culture, etc. Each county must have a County ICT director who falls under the chief information officer of Liberia or the head of the proposed ICT agency mentioned earlier.

In the area of business, I have no doubt that e-commerce will change the way business will be done in Liberia. While infrastructure is key to a successful e-commerce implementation, capacity is equally as important. With the prospect of broadband, there should be some urgency attached to building capacity. This means that business schools should begin incorporating e-commerce and m-commerce in their curricula. On the other hand, individuals, businesses, and even the government must adopt and erect web portals like ETradeLiberia.com to engage the global economy

As for the environment, the installation of a modern GIS (geographical information system) using both proprietary and open source software should be explored and possibly implemented. This GIS should be equipped with cartographic maps, satellite images, and GPS-based field observations for better management of natural resources. Also important is the installation of a GIS that contains all species of wildlife in Liberia along with another GIS initiative that uses a GIS database for eco-zoning. The Ministry of Lands, Mines, and Energy can liaise with the ICT agency mentioned earlier and the Ministry of Agriculture to accomplish this.

The GIS initiative will ensure that Liberia's mineral resources are shipped out of the country legally and that those mining are also doing so legally. Land dispute will be eradicated as the GIS and other technologies will ensure that proper land ownership is established and documented. Information on species of mammals and birds in Liberia will be stored in databases provided with the GIS, thus providing investors and developers access to information required to make development and investment decisions.

On the issue of national security, there is not much that I know of that has been put in place to illustrate that the country has or is preparing to counter modernized attacks. However, I would proffer the use Web 2.0 technologies to create interactive Web sites that will allow individuals to interact with law enforcement officers for information-sharing purposes. Again, I will recommend an information system like CENTINOL to facilitate law enforcement.

These are just a few of my many ideas on how Liberia can leverage ICT and position itself among other developing countries that are making significant strides in the global community. I am of the conviction that this transformation and the vision of a digital Liberia will soon be set in motion. And although I have not obtained a powerful position in Liberia, nor do I have the financial might to exact this vision, I am confident that someone with the resources will have the wisdom to do so. But be that as it may, I will not relinquish my quest to see a digital Liberia in full bloom. In the meantime, creating this book shall serve as my contribution. I desire that this book inspires change and provides ideas to those with the wherewithal to implement them. Oftentimes, people wish to help but are not sure where to begin. I hope this book serves as their starting point. I do not presume to know the pace at which a digital Liberia will emerge, but I do know the path to get there.

AFTERWORD

A digital Liberia will change the lives of Liberians and will move Liberia from where it once was to where it wants to be. Information and communications technology will have a great impact on Liberia, especially as it applies to education, health, business, security, government, and agriculture. A whole new generation will emerge fully knowledgeable and ready to participate in the global digital economy. Mortality rates will decrease and life expectancy rates will increase, because technology will significantly increase access to healthcare and impact the method by which it is delivered. There will be more Liberian-owned businesses as a direct result of a digital Liberia and the opportunities it will afford.

E-commerce will change the business paradigm. Farmers will have a better medium to sell their products, and Liberia's security apparatus will be fully equipped with information that can help prevent uprisings or seditious acts that once brought Africa's oldest country to its knees. Overall, a digital Liberia will be what its forefathers never dreamed possible.

In several parts of this book, I mentioned that a digital Liberia will occur only if all the stakeholders of Liberia get involved in the process. Since I am a stakeholder, one who carries a passionate desire for a digital Liberia, I too have to contribute to the process. And so I am contributing a portion of the proceeds from this book toward efforts that will lead to the infusion of technology into Liberian schools. I also volunteer to personally spearhead this process if the opportunity arises.

I believe that a digital Liberia can only succeed if we build capacity and provide the necessary infrastructure and support for it. I will continue to contribute as much as I can, and if I can.

It took me a longer time to write this book than I had initially planned, but I enjoyed every minute of it. My frequent conversations with professional Liberians as well as other Africans in the area of ICT escalated my desire for a digital Liberia. Also, my research in ICT for Development (ICT4D) has given me a broadened perspective on how we can use technology to make Africa and other developing countries better, self-sufficient, and vibrant.

I have made several recommendations and proposals in this book. One of them included the formation of a Liberian ICT authority or agency that will be solely responsible for implementing ICT in Liberia, exploring new and innovative technologies that will bring economic development and build a strong and knowledge-based society. I hope the Government of Liberia and governments in other countries that parallel Liberian's situation would heed this particular recommendation. I believe strongly that a more focused ICT entity and a dedicated and vibrant ICT and telecommunications sector can provide a lot in terms of socio-economic development.

This book is my contribution to the Liberian recovery and development process. Its intention is to create awareness and spark a debate about Liberia's future and the opportunities that lie ahead. It is my hope that all Liberians will see the big picture as I have and share my unwavering perseverance and optimism. Liberia has been given a clean slate and the chance to be great after years of poverty, illiteracy, corruption, devastation, and war. But Liberia is very resilient, as are its children. So, I will conclude this Afterword by quoting Confucius: "Our greatest glory is in never falling, but in rising every time we fall." The fall of Africa's oldest country was never its end; it was merely the beginning of its path to greater glory!

CONCLUSION

Throughout this book, I have emphasized two things that are required for a digital Liberia: connection to a submarine fiber-optic cable and a total paradigm shift from traditional approaches to those that characterize the information age. Both the former and the latter are critical to Liberia's digital revolution.

All stakeholders must participate and invest in Liberia's transformation. It is imperative that Liberia join the global community to ensure continued growth in its economy. It is crucial that Liberians at home and in the Diaspora inject themselves in this process by imparting their acquired skills, expertise, wisdom, and most importantly, support. It is necessary that Liberians adopt a new vision and embrace this relatively new culture of technology.

Just as Liberia needs a shift in paradigm—in all aspects of the Liberian society—so too should Liberians be willing to shift toward the new changes that accompany the digital revolution. For no nation should expect change when that nation itself is unready to change. It is only when this change is embraced that Liberians will achieve a digital Liberia.

ABOUT THE AUTHOR

Darren Wilkins

Darren Wilkins is an information technology professional with over twelve years of ICT experience as a technology specialist, technology architect, business analyst, writer, professional educator, and researcher. His research interests include but are not limited to information and communications technology for development (ICT4D), open source technologies, broadband connectivity, cloud computing, and e-commerce. He received an MS in computer information technology (summa cum laude), an MBA in business administration and an MS in information systems management from Hodges University in Naples, Florida.

He also holds the following ICT certifications: MCP— Microsoft Certified Professionals, MCDST—Microsoft Certified Desktop Support Technician, CCNA—Cisco Certified Network Associate, CompTIA–Linux+, CompTIA–Server+, CompTIA–Network+, and CompTIA–A+. Mr. Wilkins also holds a Florida teacher certification of education in computer science and has weaved his instructional career throughout his twelve years in the area of ICT working in both the business and the education sectors. In recent years he has focused on open source software for development (OSS4D), IT planning, policies and strategies, emerging technologies (twenty-first-century technologies), e-commerce, mobile technologies, cloud computing, submarine fiber-optic cables in Africa, and new innovations.

Mr. Wilkins is registered as a United Nations Online Volunteer and has worked on projects for the UN and other NGOs. He is the founder of the Liberian Open Source Initiative, and a founding member of the Liberian Information and Communications Technology Professionals in the United States (LICTPRO). Mr. Wilkins is a co-writer of the paper "WINWILE (Windows Interoperability with Linux in the Enterprise)" published in the *Association for Computing Machinery (ACM)* in 2004, involving the interoperability of disparate systems (Windows, Macintosh, and Linux) in the same environment. He also serves as a technology columnist for the *Liberian Daily Observer* newspaper and as the technology consultant for the B. W. Harris School in Liberia.

As far as professional affiliations goes, Mr. Wilkins is a member of the International Standards for Technology in Education (ISTE), Southwest Florida Linux User Group (SWFLUG), Atlanta Linux Enthusiasts (ALE), Computer Science Teachers Association (CSTA), and the Liberia ICT Professionals in the USA (LICTPRO).

In his spare time, Mr. Wilkins likes to read, write, travel, watch movies, play basketball and golf, and explore new and innovative technologies. He enjoys reading books on philosophy, science fiction, and politics. He spends most of his time monitoring information and communications technologies in developing countries and generating new ways to help bring economic development through ICTs. He is active in his high school alumni association in which he serves as second vice national chairman and is passionate about bringing change to the Liberian educational system through ICTs.

Mr. Wilkins currently lives in Tucker, Georgia, where he serves as the chief technology architect for Sahara Technology Solutions, LLC. He also works as an ICT consultant, researcher, instructor, columnist, blogger, and author.

GLOSSARY

A

AJAX Asynchronous JavaScript and XML. It combines several existing technologies to create dynamic Web sites that operate similarly to desktop software.
Amazon An online store and cloud-computing provider started by Jeff Bezos
Apache An open source Internet Web server
API Applications Programming Interface
Apple Computer Manufacturer of the Macintosh, and Apple computers. Also manufacturer of iPod.
APDIP Asia Pacific Development Internet Program
ATM Automated Transaction Machine
ASYCUDA Automated System of Custom Data

B

BANs Body Area Networks
Biometrics Identification and authentication via biological features
Blog Text, images, etc., maintained by a person on the Internet. Blogging began in 1998.
Bookmarking A way of marking an Internet page in order to return to it
B2B Transaction between two businesses

B2C Transaction between a business and a consumer
B2G Transaction between businesses and governments

C

C programming language High-level programming language used to develop UNIX operating system developed by DennisRitchie and Brian Kernigan at Bell Labs in 1972
CENTINOL (Central Intelligence Network of Liberia) A college research project worked on by Darren Wilkins
CPU Central Processing Unit
COBOL Common Business-Oriented Language
CIA Confidentiality, Integrity, and Availability
C2C Transaction between two consumers
CRM Customer-Relations Management

D

Database Where large information is stored, retrieved, deleted, created, etc.
DBMS Database-Management Systems; a software program that manages data within a database
Data center A location that stores servers, computers, networking equipment, people, data, etc.
DOS Disk Operating system

E

E-agriculture Electronic agriculture; agriculture involving the use of information technology
E-business Electronic business; business transaction done via information and communications technology
E-commerce Electronic commerce; commercial transaction involving the use of ICT
E-education Electronic education; using information and communications technologies in education
E-health Electronic health; providing healthcare through the use of information and communications technologies
EDI Electronic Data Interexchange

EFT Electronic File Transfer
E-TradeLIberia.com Electronic Trade Liberia; a domain name registered and owned by Darren Wilkins that was created for research purposes

F

Firewall Software or hardware that is used to prevent unauthorized access
Flickr An online community for viewing and sharing photos; www.flickr.com
FORTRAN (FORmula TRANslator) A programming language developed by IBM in 1954
FOSSFA Free Software and Open Source Foundation for Africa

G

Google Largest search engine on the Internet
GPS or Global Positioning System A collection of satellites that show locations that are displayed on computers and portable devices
GPRS General Positioning Routing System
Green ICT Green information and communications technology is an approach that involves using low-cost, low-power-consuming ICT equipment to save costs and prevent the environment from being polluted.

H

Hi-5 An online community of users who collaborate and share text, images, etc.; a social network
Hypervisor Software used in virtualization that manages several instances of an operating system on a single computer

I

ICT Information and Communications Technology
ICT4D Information Communications and Technology for Development

Informix A company that makes a DBMS (database-management software)
Interbank networks Collection of networks owned by financial institutions
IMF International Monetary Fund
Internet Collection of networks working to share resources; the global network
IIA Internet Initiative for Africa
INTERPOL INTERnational POLice, or a consortium of law enforcement officers
ISO International Organization for Standardization

J

Jambo OpenOffice An office suite originated from OpenOffice.org written for Swahili speakers that runs on Linux
JAVA An object-oriented programming language originally called OAK that was designed by Sun Microsystems and is platform independent

K

L

LANs or Local Area Networks Computer networks that are limited to a building or a location
Laubtubdina Computer Inc. (LTD) A Liberian-based computer training institution
LIBOSI LIberian Open Source Initiative
Linux Open source operating system designed by Linus Torvolds

M

Macintosh OS Operating system developed by Apple Computer, Inc.
M-business Mobile business
M-commerce Mobile commerce
Microsoft PowerPoint Presentation software provided by Microsoft Corporation packaged with MS Office
MDG Millennium Development Goals

Mozilla Firefox Open source Internet browser
MS-DOS Microsoft Disk Operating System
MySQL 5.0 An open source database language

N

NITDA Nigeria Information Technology Development Agency

O

OLTP Online Transaction Processing
OSIWA Open Source Initiative of West Africa
Open source Software A type of software designed either by an individual or a group of individuals in which the source code is made available to the public for free
Operating system The program that runs the computer. Other programs run on top of the operating system as well.
Oracle A database-management software-development company

P

PCs Personal Computers
PDA Personal Digital Assistant
POS Point-of-Sale systems
Proprietary software The opposite of open source software; these are forms of software that do not expose their code to the public. The Microsoft Office suite is an example of proprietary software.
Pro-sumer A mashup of the words producer and consumer; a new way of producing content on the Web

Q

R

RDBMS Relational-Database Management Systems
RISC Reduced-Instruction-Set Computers
RSS, or Really Simple Syndication An XML-based format used to distribute content, especially news headlines on the Internet

S

SANS Storage Area Networks
Samba An open source file- and print-sharing protocol used to allow interoperability between disparate systems
SMS, or Short Messaging Service Applications allowing the sending of short messages via mobile phones
SOAP, or Simple Object Access Protocol An XML-based messaging protocol
Social networks Web 2.0 kind of computing that involves online collaboration and sharing among individuals; Facebook, Flickr, and Hi-5 are all examples of a social network.
SQL Server Database server development by Microsoft
SDNP Sustainable Development Networking Program

T

Terminal A device usually consisting of a monitor and keyboard. It is mostly used in a server/client environment; the UNIX system is an example.
TCO Total Cost of Ownership

U

Ubuntu An African word meaning "humanity to others"
Ubuntu Linux Distribution of the Linux operating system developed by Canonical
UNCTAD United Nations Conference on Trade and Development
UNESCO United Nations Educational and Scientific Organization
United Nation ICT4D United Nation Information and Communications Technologies for Development
UNIX An operating system developed at Bell Labs

V

Virtual stores Online stores
VoIP A networking technology that allows the delivery of IP voice services over broadband

VSATs, or Very Small Aperture Terminals Small antennas or satellite dishes that allow data to be transmitted between different locations in a wide area

W

WANs Wide-Area networks
Weasua A little town in Gbarpolu, Liberia
Windows A proprietary operating system developed by Microsoft Corporation
WINWILE Windows Interoperability with Linux in the Enterprise
WSIS World Summit on Information Society
WCO World Customs Organization
WTO World Trade Organization

X

XML Extensible Marked-up Language

Y

Z

ZANACO Zambia National Commercial Bank
Zwedru The capital city of Grand Gedeh

REFERENCES

21st Century Classroom Initiative. n.d. *Monroe County: Foundation of the future.* n.d. http://www.mcschools.net/site147.php . (accessed December 11, 2009).

Abiodun Jagun. *All wrapped up! Progress on the deployment of fibre-optic submarine cables in sub-Saharan Africa.* Information Society of Africa. (2009). Http://isa.apkn.org/programme-presentations/kigali-day-2-session-8-abiodun-jagun-paper.pdf. (accessed April 11, 2010).

Abiodun Jagun. *The case for Open Access communications infrastructure in Africa: impact of international submarine cable infrastructure [SAT-3/WASC] in four African countries.* (2008). http://www.apc.org/en/system/files/APC_SAT3Briefing_20080624.pdf. (accessed March 25, 2010).

Afrique Avenir. 'IMF, World Bank Waive Liberia's Debt'. June 30, 2010. http://www.afriqueavenir.org/en/2010/06/30/imf-world bank-waive-liberia%E2%80%99s-debt/. (Accessed July 11, 2010).

Aida Opoku Mensah, Assefa Bahta and Sizo Mhlanga. "*E-Commerce Challenges in Africa: issues, constraints, opportunities.*" *United Nations Economic Commission for Africa.* 2006. http://www.uneca.org/aisi/docs/PolicyBriefs/E-commerce%20challenges%20in%20Africa.pdf (accessed October 4, 2009).

AllAfrica.com. *Govt. Allots U.S.$5M For 40 Primary Schools.* May 5, 2009. http://allafrica.com/stories/200905050678.html (accessed December 19, 2009).

Amit Singh. *Kernelthread.com.* January 2004. http://www.kernelthread.com/publications/virtualization/ (accessed April 23, 2010).

Antone Gonsalves. *Low-Power Processors: Intel Intros Atom Chips For Slimmer Netbooks. Information Week.* June 7: 23. 2010.

ASYCUDA Programme. n.d. http://www.asycuda.org/programmc.asp (Accessed March 13, 2010).

ASYCUDA. n.d.. General Benefits. http://www.asycuda.org/awbenefits.asp (Accessed March 12, 2010).

Balancing Act. *WACS consortium, Alcatel-Lucent sign contract to deploy new 14,000 km submarine cable network in West Africa.* (2009). *Internet News* 450. (accessed May 2, 2010).

Bertrand Nwankwo. *"Glo 1 Will Boost Africa's Economic Growth."* allAfrica.com. September 6, 2006. http://allafrica.com/stories/200909071164.html. (accessed August 8, 2009).

Bill Gates, Nathan Myhrvold, and Peter Rinearson. *The Road Ahead.* New York: Viking, 1995.

CDMA Development Group: http://www.cdg.org/technology/3g_1xEV-DO.asp. (accessed May 11, 2009).

Chairman King. n.d. *IBM Introduces New Netbook Software in Africa to Bridge the Digital Divide.* http://www.chairmanking.com/tag/information-technology/ (accessed June 9, 2010).

Damaria Senne. *African broadband projects hit $6.4bn.* February 29, 2008. http://www.itweb.co.za/index.php?option=com_content&view=article&id=3060&catid=. (accessed March 23, 2008).

Darren Wilkins. Liberian Daily Observer 2009(a). *The Paradox of Economic Development: Why Liberia's Infrastructural Development Initiatives Must be ICT-oriented.* May 19, 2009. http://ja-jp.facebook.com/note.php?note_id=82276628214 (accessed December 14, 2009).

Darren Wilkins. Liberian Daily Observer 2009(b). *Free Education from MIT Open Courseware* August 25, 2009. http://www.liberianobserver.com/node/1011 (accessed November 3, 2009).

Darren Wilkins. Liberian Daily Observer 2009(c). *The $1.5 Million ASYCUDA software for Customs reiterates the need for a paradigm shift in all sectors of Liberia- So, what is ASYCUDA?* July 21, 2009. Http://622.pressflex.net/news/fullstory.php/aid/17439/The_$1.5_Million_ASYCUDA_software_for_Customs_reiterates_the_need__for_a_paradigm_shift_in_all_sectors_of_Liberia-_So,_what_is_ASYCUDA_.html (accessed November 3, 2009).

Darren Wilkins. Liberian Daily Observer 2009(d). *CENTINOL A modern National Security Information System for Liberia.* September 15, 2009. http://www.liberianobserver.com/node/1579 (accessed on November 3, 2009).

Darren Wilkins. Liberian Daily Observer 2010. *An Empirical Investigation of Cloud Computing (C2) as an Option for Liberia.* April 10, 2010. http://www.liberianobserver.com/node/5665 (accessed on June 3, 2010).

Darren Wilkins. *Plugging into the "Cloud"- How Cloud Computing can enhance ICT penetration in Liberia.* Darren Wilkins's Blog. October 26, 2009. http://darrenwilkins.wordpress.com/2009/10/26/4/ (accessed April 4, 2010).

Dennis Onwuegbu. *Changing Face of Mobile Telephony from Numbers to Values: The value system is changing in the telecoms industry in Africa.* n.d. http://ittelecomdigest.com/cover.htm. (accessed January 23, 2010).

Deshpande. A, & Riehle, D. *The Total Growth of Open Source.* (2008). http://dirkriehle.com/wp-content/uploads/2008/03/oss-2008-total-growth-final-web.pdf . (accessed March 23, 2010).

Dow Jones Newswires. *Maroc Telecom to link Morocco, West Africa via fibre: First phase of backbone project reportedly 60% complete.* February 25, 2010. http://www.totaltele.com/view.aspx?ID=453467. (accessed February 28, 2010).

Eduardo. Da Costa. *Global E-Commerce Strategies For Small Business.* Cambridge, Massachusetts: MIT. 2001.

Ecobank. n.d. *VISA Card.* http://www.davidajao.com/blog/2007/07/11/ecobank-ghana-visa-gold-credit-card/. (accessed June 13, 2009).

E-commerce Journal. *Splash and More Magic Solutions launch mobile payments in Sierra Leone.* (n.d.). Http://www.ecommerce-journal.com/news/18236_splash_and_moremagic_solutions_launch_mobile_payments_in_sierra_leone. (accessed November 27, 2009).

Ecommerce Journal. *The International Bank of Liberia Has Launched SMS Service.* (2008). Http://www.ecommerce-journal.com/news/the_international_bank_of_liberia_has_launched_sms_service. (accessed June 13, 2009).

Economist Intelligence Unit. *Overview of E-commerce in Nigeria.* April 4. 2006. http://globaltechforum.eiu.com/index.asp?categoryid=&channelid=&doc_id=8403&layout=rich_story&search=nigerian. (accessed November 27, 2009).

Efem Nkanga. *"Adenuga: 'Glo 1' Will Bring New Prosperity to Africa"* This Day Special Release. September 7, 2009. http://www.nigerianmuse.com/20090909123110zg/projects/TelecomProject/adenuga-glo-1-will-bring-new-prosperity-to-africa-as-glo-1-submarine-cable-lands-in-lagos. (accessed November 12, 2009).

Efraim Turban, Jae Kyu Lee, Dave King, Judy McKay and Peter Marshall. *Electronic Commerce 2008: A Managerial Perspective.* Upper Saddle River. Pearson Education, Inc. 2008.

eGovernment for Development. *What is eGovernment?* (2010). http://www.egov4dev.org/index.shtml (accessed January 18, 2010).

Erick Schonfeld. *Techcrunch -IBM's Blue Cloud is Web Computing By Another Name.* November 5, 2007. http://techcrunch.com/2007/11/15/ibms-blue-cloud-is-web-computng-by-another-name/ (accessed May 6, 2010).

ETranzact. *ETranzact revolutionizes E-commerce in Ghana.* (March 17, 2009). http://ghanabusinessnews.com/2009/03/17/etranzact-revolutionizes-E-commerce-in-ghana/. (accessed November 27, 2009).

Fiber for Africa. *Who are SAT-3's Members?* (2009). http://fibreforafrica.net/main.shtml?x=4080673&als[MYALIAS6]=Who are SAT3's members?&als[select]=4018621. (accessed May 2010).

Garneo Cephas. *The Vision.* July 31, 2009. http://thevisionliberia.org/index.php?option=com_content&view=article&id=131:ecobank-introduces-bankers-visa-card-in-liberia&catid=32:business&Itemid=47 (Accessed August 3, 2009).

Gary McGraw. *Software Security: Building Security In.* Boston: Pearson Education Inc., 2009.

George Kennedy. *GOL Brings in US$1.5M Software.* July 15, 2009. Http://622.pressflex.net/news/fullstory.php/aid/17367/GOL_Brings_in_US$1.5M_Software_.html (accessed March 12, 2010).

Georgia Tech Press Release. *Re-inventing Telecom Technology in Liberia.* April 18, 2007. Http://www.gatech.edu/newsroom/release.html?id=1343. (accessed April 1, 2009).

Glossary of Terms. *MaIN OnE*: http://www.voxtelecom.co.za/investor/glossary.asp. (accessed May 12, 2010).

California4gore. California Draft. *Gore 08: Save America. Save the world.* (n.d.). http://www.california4gore.org/Al_Gore_Invented_The_Internet.html. (accessed April 4, 2010).

Government, Center of Technology in. "The Future of E-Government: A Working Definition of E-Government." *Center of Technology in Government- State Univeristy of New York, University at Albany.* n.d. http://www.ctg.albany.edu/publications/reports/future_of_egov?chapter=2 (accessed February 18, 2010).

Hsinchun Chen and Roslin. V. Hauk. Coplink: "A case of intelligent analysis and knowledge management."(1999). Proceedings of the 20th International Conference on Information Systems. ACM, pp. 15–28.

H. Chen, H. Atabakhsh, D. Zeng, J. Schroeder, T. Petersen, D. Casey, M. Chen, Y. Xiang, D. Daspit, and S. Fu. Nandiraju. (2002). "*COP:* Visualization and collaboration for law enforcement." Proceedings of the 2002 annual national conference on Digital government research. ACM, pp. 1-7.

H. Willie. *Liberia Faces Serious Capacity Challenge.* March 28, 2009. http://www.liberiawebs.com/index.php?option=com_content&view=article&id=1238:liberia-faces-serious-capacity-challenge-&catid=125:finance&Itemid=371. (accessed April 2, 2009).

Indian Institute of Technology. http://netlab.cs.iitm.ernet.in/cs648/2009/assignment2/ma08m012.pdf. (accessed November 7, 2009).

Innovation Africa. *IBM and Cloud Computing in Africa.* June 18, 2010. Http://www.innovationafrica.org/2010/06/ibm-and-cloud-computing-in-africa/ (accessed June 22, 2010).

International Monetary Fund, (2008). *Liberia: Poverty Reduction Strategy Paper.* (2008). Http://www.imf.org/external/pubs/ft/scr/2007/cr07313.pdf. (accessed November 12, 2009).

IT News Africa. *Technology helps ANC win elections.* April 30, 2009. http://www.itnewsafrica.com/?p=2572 (accessed November 22, 2009).

ITU. *World Telecommunication/ICT Indicators Database 2010.* Geneva: ITU.

J. D. Biersdorfer. *Netbooks: the missing manual- The book that should have been in the box.* Sebastopol: O'Reilly Media, Inc., 2009.

J. K. Roberts. *Suspected Terrorists Legal Battle Begins.* http://www.newdemocratnews.com/story.php?record_id=1528&sub=14 (accessed February 12, 2009).

John K. Waters. CIO.com. n.d. http://www.cio.com/article/40701/Virtualization_Definition_and_Solutions (accessed May 3, 2010).

Katherine N. Andrews, Brandon L. Hunt, and William J. Durch. *Post-Conflict Border and Peace Operations.* Report from the project on Rule of Law in Post-Conflict Settings.(2007). The Henry L. Stimson Center. (accessed August 11, 2009).

John Lamb. *The Greening of IT: How Companies Can Make a Difference for the Environment.* Indianapolis: IBM Press, 2009.

Liberia, Republic of. "*Telecommunications Act 2007.*" Ministry of Post and Telecommunications. 2007. http://www.mopt.gov.lr/doc/telecom_act_2007.pdf (accessed August 7, 2009).

Liberian Daily Observer. *Central Bank Prepares to Develop Stock Market. December 31,* 2009. http://www.liberianobserver.com/node/3749 (accessed January 4, 2010).

LIBTELCO. *LIBTELCO-Profile.* 2009. http://www.libtelco.com.lr/profile.html (accessed January 17, 2010).

Lionel Bernard. *A Case for Privatization of Telecommunications in Liberia.* The Perspective Online Newsmagazine. *(*July 16, 2004). http://www.theperspective.org/2004/july/privatizationoftelecom.html. (accessed June 23, 2009).

Main One Cable Company. *Project Room.* http://www.mainonecable.com/projectroom.php. (accessed April 21, 2010).

Main One. *Main One submarine cable project inches near completion.* Sub Telecom Forum. February 1. http://www.subtelforum.com/articles/?p=2035. (accessed April 21, 2010).

Michael Best & Dhanaraj Thakur. *The telecommunications policy process in post-conflict developing countries – the case of Liberia.* (2008) http://mikeb.inta.gatech.edu/uploads/papers/TPRC.2008.Liberia.pdf (accessed February 11, 2009).

Michael Best, Dhanaraj Thakur, Kipp Jones, Illenin Kondo, Dhanaraj Thakur, Edem Wornyo, and Calvin Yu. Post-Conflict Communications: The Case of Liberia. (2007). *Communications of the ACM*, 50(10), 33-39.

Michael Kpayili. Liberia: *President Sirleaf Strategizes Fight Against Corruption.* Liberia Mandigo Association of New York. June 8, 2010.

http://limany.org/index.php?option=com_content&view=article&id=418:liberia-president-sirleaf-strategizes-fight-against-corruption (accessed June 9, 2010).

Michael Malakata. *Electronic Banking prepares the way for e-commerce in Zambia*. http://www.ftpiicd.org/iconnect/information and communications technology (ICT)4D_Livelihoods/ZM_Livelihoods_EN.pdf. (accessed March 11, 2009).

Michael Miller. *Cloud Computing: Web-Based Applications That Change the Way You Work and Collaborate Online*. Indianapolis: Que Publishing, 2008.

Michael Ruddy. *"Undersea Cable Markets and the Developing World." www.terabitconsulting.com*. May 2007. http://www.terabitconsulting.com/public/downloads/Terabit%20Consulting%20-%20Undersea%20Cable%20Markets%20and%20the%20Developing%20World.htm (accessed September 5, 2009).

Ministere de la Sante et de la Prevention. *www.sante.gouv.sn*

Ministério das Finanças. http://www.minfin.cv (accessed December 2, 2009).

MIT OpenCourseWare. *Now Available at the American University in Cairo Hosted and Managed by University Academic Computing Technologies*. http://mitocw.aucegypt.edu/. (accessed April 2, 2009).

Mona Abdulla. *The Main Components of E-Commerce*. (n.d.). http://ezinearticles.com/?The-Main-Components-of-E-Commerce&id=2298719. (accessed on November 21, 2009).

Nate Anderson. *An Introduction to IPTV: Television is changing*. March 12, 2006. http://arstechnica.com/business/news/2006/03/iptv.ars (Accessed May 12, 2010).

National Association of Broadcasters. n.d. *Stop Radio Tax*. http://www.noperformancetax.org/Radio%20at%20Risk (accessed February 12, 2010).

Nations, United. *"United Nations e-Government Survey 2008."* United Nations Public Administration Network. 2008. http://unpan1.un.org/intradoc/groups/public/documents/un/unpan028607.pdf (accessed January 6, 2010).

Omojola Biodun. *Out of Digital Wilderness, Africa Telecom Today.* April 4. 2009. http://www.africatoday.com/cgibin/public.cgi?sub=news&action=one&cat=180&id=1731. (accessed January 12, 2010).

OpenCourseware, MIT. *"MIT OpenCourseware - Our Story." MIT OpenCourseware- Massachusettes Institute of Technology.* n.d. http://www.ocw.nur.ac.rw/OcwWeb/Global/AboutOCW/our-story.htm (accessed December 20, 2009).

OpenOffice.org. *klnX.* December 24, 2004. http://www.o.ne.tz/ (accessed March 23, 2010).

Orange. Press Release. *"ACE (Africa Coast to Europe) submarine cable welcomes new members."* December 1, 2009 http://www.orange.com/en_EN/press/press_releases/att00014018/PR_ACE_en_011209.pdf. (accessed January 9, 2010).

Overseas Development Institute. *Background Information on Poverty Reduction Strategy Papers and the Water Sector. n.*d. http://www.odi.org.uk/resources/download/3216.pdf. (accessed November 12, 2009).

Panwhanpen. Panwhanpen comments on *"3-Yrs of Unity party Rule Have Change Things? Nation's Address to Leg. Speaks of Progress & Challenges."* The Panwhanpen blog. Comment posted on January 27, 2009. http://panwhanpen.blogspot.com/2009/01/3-yrs-of-unity-party-rule-have-change.html. (accessed April 17, 2009).

Rashid Mahmood. *New Ways to Learn: Online education and other innovative initiatives are reducing costs and increasing accessibility while helping to meet the needs of students and business.* American Chamber of Commerce in Egypt http://www.amcham.org.eg/Publications/BusinessMonthly/November%2009/coverstory.asp. (accessed November 12, 2009).

Rebekah Peacock. *Damaged cable causes Internet blackout in four West African countries.* July 29, 2009. http://opennet.net/blog/2009/07/damaged-cable-causes-internet-blackout-four-west-african-countries. (accessed November 12, 2009).

Red Herring Magazine. Getting E-commerce to Africa. March 1, 2001. http://www.redherring.com/Home/9506. (accessed February 12, 2009).

Republic of Liberia. *"Poverty Reduction Strategy."*. Lift Liberia. 2006. http://www.liftliberia.gov.lr/doc/prs.pdf (accessed March 14, 2010).

Reuters. 2009. *UPDATE 1-Telecoms operators sign African undersea cable deal.* April 8, 2009. http://www.reuters.com/article/idUSL896481320090408 (accessed May 11, 2009).

Rich Miller. *How Google's Ocean Power Would Work.* (September 25, 2008). http://www.datacenterknowledge.com/archives/2008/09/25/how-googles-ocean-power-would-work/ (accessed March 18, 2010).

Rob Katz. *Cell Phone Banking and the BOP: Wizzit.* June 19, 2006. http://www.nextbillion.net/blog/2006/06/19/cell-phone-banking-and-the-bop-wizzit (accessed May 9, 2009).

Robert W. Hahn. *Government Policy toward Open Source Software.* Washington: Brookings Institute Press, 2002.

Russell Southwood. Liberia: Four mobile companies bring lowest prices in West Africa. *Balancing Act.* (2007). Http://www.balancingact- africa.com/news/back/balancing-act 352.html

Russell Southwood. *Revealed: The SAT3 Consortium's Shareholder Agreement They Don't Want You To See.* (2006). http://mybroadband.co.za/nephp/3787.html (accessed February 11, 2010).

SAT-3\WASC\SAFE. 2009. http://www.safe-sat3.co.za/ (accessed March 2009).

Song Steve. *Sub-Saharan Africa Submarine Cables in 2010.* September 25, 2008 http://manypossibilities.net/2008/09/sub-saharan-africa-submarine-cables-in-2010-2/comment-page-1/. (accessed January 21, 2008).

Staff Writer. *SAT-3 competitor ACE gets new members.* December 3, 2009. http://mybroadband.co.za/news/Telecoms/10718.html. (accessed April 11, 2010).

Stephen M. Mutula. *Digital divide and economic development: Case study of sub-Saharan Africa*. Department of Library and Information Studies, University of Botswana, Gaborone, Botswana. The Electronic Library (2008). 26 (4):468–489.

Sylvia Ramsey. Sylviaramseyauthor.com blog. *Piercing the Underground—Cyber Crime a Real Threat by Foreign Entities.* April 12. 2009. http://sylviaramseyauthor.com/blog/2009/04/12/piercing-the-underground-cyber-crime-a-real-threat-by-foreign-entities/ (accessed February 14, 2010).

T. Newcombe. *Game-Changer? Public CIO-Technology Leadership in the Public Sector.* (January 2010). Issue 6|Vol. 7. E-Republic

Telecommunications, Ministry of Post and. *National ICT Policy Draft.* (2009). http://www.mopt.gov.lr/press.php?news_id=18 (accessed January 22, 2010).

Telecommunications, Ministry of Post and. *Feasibility Study Of the Economic and Financial Viability Of Undersea Fiber Connectivity To the Mano River Union Countries (Liberia, Guinea & Sierra Leone)."* (October 13, 2008) http://www.mopt.gov.lr/doc/Feasibility_Study_MRU_SAT3_101308.pdf (accessed April 2010).

TeleGeography. *Submarine Cable Map.* 2010. http://www.telegeography.com/product-info/map_cable/index.php. (accessed April 11, 2010).

Terrence Creamer. *Competing submarine cable projects creating redundancy risk.* June 29, 2007. http://www.engineeringnews.co.za/article/competing-submarine-cable-projects-creating-redundancy-risk-2007-06-29. (accessed June 7, 2009).

The Economist. "Liberia's feisty president: *Another round for Africa's Iron Lady. A woman's work is never done.* May 20, 2010. http://www.

economist.com/world/middleeast-africa/displaystory.cfm?story_id=16168384. (accessed June 1, 2010).

Tiemann, Michael. *Opensource.Org.* November 1, 2009. http://www.opensource.org/files/OSS-2010.pdf (accessed March 4, 2010).

TLCAfrica. Technology. June 5, 2010. http://www.tlcafrica.com/technology.htm (accessed June 7, 2010).

TLC Africa. "Policy Framework for Connecting Liberia to Undersea Fiber Cable. Ministry of Post and Telecommunications (2010) http://www.tlcafrica.com/technology_policy_framework_connecting_liberia_to_underwater_fiber_cable.htm (accessed May 19, 2010).

Trade Logistic Branch. n.d. http://r0.unctad.org/ttl/docs-

brochure/UNCTAD_SDTE_TLB_MISC_2007_9_3_VOLETS.pdf (accessed March 11, 2010).

UNCTAD. 2001. E-commerce and Development Report. United Nations, Geneva.

UNCTAD. 2004. E-commerce and Development Report. United Nations, Geneva.

UNCTAD. 2009. E-commerce and Development Report. United Nations, Geneva.

UNCTAD. n.d. ASYCUDA: Automated Systems for Customs Data-Technology. http://www.asycuda.org/awtechnology.asp (Accessed March 13, 2010).

Union, International Telecommunications. *Measuring the Information Society - The ICT Development Index.* March 19, 2009. http://www.itu.int/ITU-D/ict/publications/idi/2009/index.html (accessed December 11, 2009).

United Nations Development Program. n.d. Information Communications and Technologies For Development. http://ictd.undp.org/it4dev/docs/about_undp.html (accessed February 12, 2010).

United Nations Development Programme. n.d. Information and Communications Unit. http://mirror.undp.org/liberia/ict.htm. (accessed March 26, 2009)

UNPAN. 2008. *UN E-Government Survey 2008.*

Vargas, Jose Antonio. *The Washington Post.* November 28, 2008. http://email00.secureserver.net/webmail.php?login=1&ssl=0&gzipjs=1 (accessed December 22, 2009).

Wikipedia, *ACE (cable system)*. 2009. http://en.wikipedia.org/wiki/ACE cable system (accessed December 28, 2009).

Wikipedia, *Liberia—Economy.* n.d.. http://en.wikipedia.org/wiki/Liberia#Economy (accessed November 11, 2009).

Wikipedia, *Poverty Reduction Strategy Paper.* n.d. Answers.com. http://www.answers.com/topic/poverty-reduction-strategy-paper. (accessed April 11, 2009).

Wikipedia. *Total Cost of Ownership (TCO).* May 14, 2010. http://en.wikipedia.org/wiki/Total_cost_of_ownership (accessed June 11, 2010).

William Minter. *Africa Files.* October 27, 2008. http://www.africafiles.org/article.asp?ID=22166 (accessed February 11, 2010).

Wisconsin. University of Wisconsin Distance Learning. n.d. http://distancelearning.wisconsin.edu/technologies.htm (Accessed January 17, 2010).

INDEX

A

B

C

D

E

I

J

K

L

M

N

Q

R

S

T

U

V

W

X

Y

Z